The IMA Volumes in Mathematics and its Applications

Volume 54

Series Editors
Avner Friedman Willard Miller, Jr.

Institute for Mathematics and its Applications

IMA

The **Institute for Mathematics and its Applications** was established by a grant from the National Science Foundation to the University of Minnesota in 1982. The IMA seeks to encourage the development and study of fresh mathematical concepts and questions of concern to the other sciences by bringing together mathematicians and scientists from diverse fields in an atmosphere that will stimulate discussion and collaboration.

The IMA Volumes are intended to involve the broader scientific community in this process.

Avner Friedman, Director
Willard Miller, Jr., Associate Director

* * * * * * * * * *

IMA ANNUAL PROGRAMS

1982–1983 **Statistical and Continuum Approaches to Phase Transition**
1983–1984 **Mathematical Models for the Economics of Decentralized Resource Allocation**
1984–1985 **Continuum Physics and Partial Differential Equations**
1985–1986 **Stochastic Differential Equations and Their Applications**
1986–1987 **Scientific Computation**
1987–1988 **Applied Combinatorics**
1988–1989 **Nonlinear Waves**
1989–1990 **Dynamical Systems and Their Applications**
1990–1991 **Phase Transitions and Free Boundaries**
1991–1992 **Applied Linear Algebra**
1992–1993 **Control Theory and its Applications**
1993–1994 **Emerging Applications of Probability**
1994–1995 **Waves and Scattering**

IMA SUMMER PROGRAMS

1987 **Robotics**
1988 **Signal Processing**
1989 **Robustness, Diagnostics, Computing and Graphics in Statistics**
1990 **Radar and Sonar (June 18 - June 29)**
New Directions in Time Series Analysis (July 2 - July 27)
1991 **Semiconductors**
1992 **Environmental Studies: Mathematical, Computational, and Statistical Analysis**
1993 **Modeling, Mesh Generation, and Adaptive Numerical Methods for Partial Differential Equations**

* * * * * * * * * *

SPRINGER LECTURE NOTES FROM THE IMA:

The Mathematics and Physics of Disordered Media
Editors: Barry Hughes and Barry Ninham
(Lecture Notes in Math., Volume 1035, 1983)

Orienting Polymers
Editor: J.L. Ericksen
(Lecture Notes in Math., Volume 1063, 1984)

New Perspectives in Thermodynamics
Editor: James Serrin
(Springer-Verlag, 1986)

Models of Economic Dynamics
Editor: Hugo Sonnenschein
(Lecture Notes in Econ., Volume 264, 1986)

David Kinderlehrer Richard James
Mitchell Luskin Jerry L. Ericksen
Editors

Microstructure and Phase Transition

With 60 Illustrations

Springer-Verlag
New York Berlin Heidelberg London Paris
Tokyo Hong Kong Barcelona Budapest

David Kinderlehrer
Department of Mathematics
Carnegie Mellon University
Pittsburgh, PA 15213
USA

Richard James
Department of Aerospace Engineering
and Mechanics
University of Minnesota
Minneapolis, MN 55455
USA

Mitchell Luskin
School of Mathematics
University of Minnesota
Minneapolis, MN 55455
USA

Jerry L. Ericksen
5378 Buckskin Bob Drive
Florence, OR 97439
USA

Series Editors:
Avner Friedman
Willard Miller, Jr.
Institute for Mathematics and
its Applications
University of Minnesota
Minneapolis, MN 55455
USA

Mathematics Subject Classifications (1991): 35J20, 35J85, 35K55, 35K85, 35Q35, 34R05, 35R35, 93E05, 65N55, 49J10, 49J20, 49J40

Library of Congress Cataloging-in-Publication Data
Microstructure and phase transition / David Kinderlehrer . . . [et al.], editors.
p. cm. – (The IMA volumes in mathematics and its applications ; v. 54)
Includes bibliographical references.
ISBN 0-387-94112-6
1. Microstructure – Mathematical models – Congresses. 2. Variational inequalities (Mathematics) – Congresses. 3. Phase rule and equilibrium – Congresses. I. Kinderlehrer, David. II. Series.
TA407.M53 1994
620.1′1299 – dc20 93-5144

Printed on acid-free paper.

Production managed by Bill Imbornoni; manufacturing supervised by Jacqui Ashri.
Camera-ready copy prepared by the IMA.
Printed and bound by Edwards Brothers, Inc., Ann Arbor, MI.
Printed in the United States of America.

9 8 7 6 5 4 3 2 1

ISBN 0-387-94112-6 Springer-Verlag New York Berlin Heidelberg
ISBN 3-540-94112-6 Springer-Verlag Berlin Heidelberg New York

The IMA Volumes
in Mathematics and its Applications

Current Volumes:

Volume 1: Homogenization and Effective Moduli of Materials and Media
Editors: Jerry Ericksen, David Kinderlehrer, Robert Kohn, J.-L. Lions

Volume 2: Oscillation Theory, Computation, and Methods of Compensated Compactness
Editors: Constantine Dafermos, Jerry Ericksen,
David Kinderlehrer, Marshall Slemrod

Volume 3: Metastability and Incompletely Posed Problems
Editors: Stuart Antman, Jerry Ericksen, David Kinderlehrer, Ingo Müller

Volume 4: Dynamical Problems in Continuum Physics
Editors: Jerry Bona, Constantine Dafermos, Jerry Ericksen,
David Kinderlehrer

Volume 5: Theory and Applications of Liquid Crystals
Editors: Jerry Ericksen and David Kinderlehrer

Volume 6: Amorphous Polymers and Non-Newtonian Fluids
Editors: Constantine Dafermos, Jerry Ericksen, David Kinderlehrer

Volume 7: Random Media
Editor: George Papanicolaou

Volume 8: Percolation Theory and Ergodic Theory of Infinite Particle Systems
Editor: Harry Kesten

Volume 9: Hydrodynamic Behavior and Interacting Particle Systems
Editor: George Papanicolaou

Volume 10: Stochastic Differential Systems, Stochastic Control Theory and Applications
Editors: Wendell Fleming and Pierre-Louis Lions

Volume 11: Numerical Simulation in Oil Recovery
Editor: Mary Fanett Wheeler

Volume 12: Computational Fluid Dynamics and Reacting Gas Flows
Editors: Bjorn Engquist, M. Luskin, Andrew Majda

Volume 13: Numerical Algorithms for Parallel Computer Architectures
Editor: Martin H. Schultz

Volume 14: Mathematical Aspects of Scientific Software
Editor: J.R. Rice

Volume 15: Mathematical Frontiers in Computational Chemical Physics
Editor: D. Truhlar

Volume 16: Mathematics in Industrial Problems
by Avner Friedman

Volume 17: Applications of Combinatorics and Graph Theory to the Biological and Social Sciences
Editor: Fred Roberts

Volume 18: q-Series and Partitions
Editor: Dennis Stanton

Volume 19: Invariant Theory and Tableaux
Editor: Dennis Stanton

Volume 20: Coding Theory and Design Theory Part I: Coding Theory
Editor: Dijen Ray-Chaudhuri

Volume 21: Coding Theory and Design Theory Part II: Design Theory
Editor: Dijen Ray-Chaudhuri

Volume 22: Signal Processing: Part I - Signal Processing Theory
Editors: L. Auslander, F.A. Grünbaum, J.W. Helton, T. Kailath, P. Khargonekar and S. Mitter

Volume 23: Signal Processing: Part II - Control Theory and Applications of Signal Processing
Editors: L. Auslander, F.A. Grünbaum, J.W. Helton, T. Kailath, P. Khargonekar and S. Mitter

Volume 24: Mathematics in Industrial Problems, Part 2
by Avner Friedman

Volume 25: Solitons in Physics, Mathematics, and Nonlinear Optics
Editors: Peter J. Olver and David H. Sattinger

Volume 26: Two Phase Flows and Waves
Editors: Daniel D. Joseph and David G. Schaeffer

Volume 27: Nonlinear Evolution Equations that Change Type
Editors: Barbara Lee Keyfitz and Michael Shearer

Volume 28: Computer Aided Proofs in Analysis
Editors: Kenneth Meyer and Dieter Schmidt

Volume 29: Multidimensional Hyperbolic Problems and Computations
Editors: Andrew Majda and Jim Glimm

Volume 30: Microlocal Analysis and Nonlinear Waves
Editors: Michael Beals, R. Melrose and J. Rauch

Volume 31: Mathematics in Industrial Problems, Part 3
by Avner Friedman

Volume 32: Radar and Sonar, Part 1
by Richard Blahut, Willard Miller, Jr. and Calvin Wilcox

Volume 33: Directions in Robust Statistics and Diagnostics: Part I
Editors: Werner A. Stahel and Sanford Weisberg

Volume 34: Directions in Robust Statistics and Diagnostics: Part II
Editors: Werner A. Stahel and Sanford Weisberg

Volume 35: Dynamical Issues in Combustion Theory
Editors: P. Fife, A. Liñán and F.A. Williams

Volume 36: Computing and Graphics in Statistics
Editors: Andreas Buja and Paul Tukey

Volume 37: Patterns and Dynamics in Reactive Media
Editors: Harry Swinney, Gus Aris and Don Aronson

Volume 38: Mathematics in Industrial Problems, Part 4
by Avner Friedman

Volume 39: Radar and Sonar, Part II
Editors: F. Alberto Grünbaum, Marvin Bernfeld and Richard E. Blahut

Volume 40: Nonlinear Phenomena in Atmospheric and Oceanic Sciences
Editors: G. F. Carnevale and R.T. Pierrehumbert

Volume 41: Chaotic Processes in the Geological Sciences
Editor: David A. Yuen

Volume 42: Partial Differential Equations with Minimal Smoothness and Applications
Editors: B. Dahlberg, E. Fabes, R. Fefferman, D. Jerison, C. Kenig and J. Pipher

Volume 43: On the Evolution of Phase Boundaries
Editors: M.E. Gurtin and G. McFadden

Volume 44: Twist Mappings and Their Applications
Editor: R. McGehee and K.R. Meyer

Volume 45: New Directions in Time Series Analysis, Part I
Editors: David Brillinger, Peter Caines, John Geweke, Emanuel Parzen, Murray Rosenblatt, and Murad S. Taqqu

Volume 46: New Directions in Time Series Analysis, Part II
Editors: David Brillinger, Peter Caines, John Geweke, Emanuel Parzen, Murray Rosenblatt, and Murad S. Taqqu

Volume 47: Degenerate Diffusions
Editors: W.-M. Ni, L.A. Peletier, J.-L. Vazquez

Volume 48: Linear Algebra, Markov Chains and Queing Models
Editors: Carl D. Meyer and Robert J. Plemmons

Volume 49: Mathematics in Industrial Problems, Part 5
by Avner Friedman

Volume 50: Combinatorial and Graph-Theoretical Problems in Linear Algebra
Editors: Richard Brualdi, Shmuel Friedland and Victor Klee

Volume 51: Statistical Thermodynamics and Differential Geometry of Microstructured Material
Editors: H. Ted Davis and Johannes C.C. Nitsche

Volume 52: Shock Induced Transitions and Phase Structures
Editors: J.E. Dunn, Roger Fosdick and Marshall Slemrod

Volume 53: Variational and Free Boundary Problems
Editors: Avner Friedman and Joel Spruck

Volume 54: Microstructure and Phase Transitions
Editors: David Kinderlehrer, Richard James, Mitchell Luskin and Jerry L. Ericksen

Forthcoming Volumes:

Dynamical Systems and Their Applications

Dynamical Theories of Turbulence in Fluid Flows

Phase Transitions and Free Boundaries

Free Boundaries in Viscous Flows

Summer Program: *Semiconductors*

Semiconductors (2 volumes)

Applied Linear Algebra

Sparse Matrix Computations: Graph Theory Issues and Algorithms

Iterative Methods for Sparse and Structured Problems

Linear Algebra for Signal Processing

Linear Algebra for Control Theory

Summer Program: *Environmental Studies*

Environmental Studies

FOREWORD

This IMA Volume in Mathematics and its Applications

MICROSTRUCTURE AND PHASE TRANSITION

is based on the proceedings of a workshop which was an integral part of the 1990-91 IMA program on "Phase Transitions and Free Boundaries." We thank R. Fosdick, M.E. Gurtin, W.-M. Ni and L.A. Peletier for organizing the year-long program and, especially, D. Kinderlehrer, R. James, M. Luskin and J. Ericksen for organizing the meeting and editing these proceedings.

We also take this opportunity to thank those agencies whose financial support made the workshop possible: the Army Research Office, and the National Science Foundation.

Avner Friedman

Willard Miller, Jr.

PREFACE

Much of our traditional knowledge of materials and processes is achieved by observation and analysis of small departures from equilibrium. Many materials, especially modern alloys, ceramics, and their composites, experience not only larger but more dramatic changes, such as the occurrence of phase transitions and the creation of defect structures, when viewed at the microscopic scale. How is this observed, how can it be interpreted, and how does it influence macroscopic behavior? These are the principle concerns of this volume, which constitutes the proceedings of an IMA workshop dedicated to these issues.

Crystalline solids, of which alloys and ceramics are particular examples, are often idealized as materials with a high degree of configurational order. The configurational order may be viewed as a constraint which, when compromised, gives rise to a defect structure in the material. Nonlinear theory presents the opportunity to understand and control this behavior, whose mathematical context lies in the study of constrained variational problems and systems of differential equations which may be unstable or incompletely posed. The reader will discover in this volume current progress in some of the powerful new methods introduced to study these questions as well as recent experimental results.

There are many potentially relevant technological applications implicated here. Such applications include shape memory materials, nonlinear dielectrics and ferroelectrics, magnetostrictive materials, composites, or other materials fabricated or devised to promote special properties.

We would like to thank the authors for preparing manuscripts accessible to a wide audience while maintaining a high level of scientific content. We would also like to thank Avner Friedman and Willard Miller of the IMA and the program coordinator Morton Gurtin for the opportunity to bring these issues before the applied mathematics community.

David Kinderlehrer

Richard James

Mitchell Luskin

Jerry L. Ericksen

CONTENTS

THE MICROSTRUCTURE OF MARTENSITE AND ITS IMPLICATIONS FOR THE SHAPE-MEMORY EFFECT

KAUSHIK BHATTACHARYA*

1. Introduction. Crystalline solids undergoing martensitic phase transformation display fine scale microstructure. The microstructure plays an important role in determining the macroscopic properties of these solids. The aim of this investigation is to understand what role the lattice parameters of the material play in determining the microstructure and consequently the macroscopic properties. The lattice parameters measure the size and shape of the crystal lattices and may be determined experimentally. In particular, the aim here is to determine if the lattice parameters of the technologically significant shape-memory alloys are special in any respect.

Martensitic transformation is a temperature induced, first order, diffusionless transformation. Thus, a material undergoing this transformation has one crystal structure above a critical temperature and another below the critical temperature. As the temperature is changed, the change from one structure to another is discontinuous, purely structural and diffusionless. The high temperature phase is called austenite and the low temperature phase is called martensite. Often the lattice of the austenite has greater symmetry than that of the martensite. Thus, it is possible to have variants of martensite – variants are identical lattices of martensite which are oriented differently with respect to the austenite lattice. The variants arrange themselves in fine scale microstructure in order to satisfy imposed boundary conditions. These microstructures are coherent and free of macroscopic internal stress. The martensite can deform by rearranging the variants into new microstructures. The lattice parameters determine the coherent, stress-free microstructures that a material can form and thereby influence the macroscopic properties. Certain materials undergoing this transformation display the shape-memory effect. This is a phenomenon wherein an apparently plastically deformed specimen recovers all strain when heated to above a critical temperature.

I use a continuum model based on nonlinear thermoelasticity to study this problem. This model is similar to the one described by Professor J.L. Ericksen at this workshop. It is derived by linking the displacement of the atoms to the macroscopic deformation using the Cauchy–Born hypothesis [24, 69]. The continuum model is presented in §2 and the origin of the fine scale microstructure is explained. In order to understand the effect of lattice parameters on the microstructure, I study the wedge-like microstructure in §3. I explain the shape-memory effect and study the problem of self-accommodation in §4. The methods used are very similar to those used by Professor J.M. Ball amongst others at this workshop.

*Department of Aerospace Engineering and Mechanics, University of Minnesota, 107 Akerman Hall, Minneapolis, MN 55455. Current address: Courant Institute of Mathematical Sciences, 251 Mercer Street, New York, NY 10012

The wedge-like microstructure is commonly seen in materials that undergo a reversible martensitic transformation, especially in shape-memory alloys [1, 2, 16, 48–54, 63, 64]. This microstructure provides an easy mechanism for the initiation of transformation and is important for thermoelasticity and reversibility of the transformation. The main result of §3 is that only very special materials, those whose lattice parameters satisfy certain strict condition, can form the wedge-like microstructure in a stress-free and coherent manner.

A key feature of the shape-memory effect is that during cooling, the transformation from the austenite to the martensite takes place without any macroscopic change in shape. At the atomic level there is a structural transformation. However, the variants of martensite arrange themselves in such a microstructure that there is no macroscopic change in shape. This is known as self-accommodation. Apart from being an inherent part of the shape-memory effect, self-accommodation is important for the reversibility of transformation in polycrystals and for easy nucleation of martensite during cooling. The main result of §4 is summarized in Table 4.1. If the symmetry of the austenite lattice is cubic, we have self-accommodation if and only if the transformation is volume preserving. If the symmetry of the austenite is not cubic, the material has to satisfy additional, extremely strict conditions for self-accommodation. It is interesting to note that even though it is the martensite that forms the microstructure, the results depend only on the symmetry of the austenite.

The results of §3 and §4 are compared with experimental observations in §5. §6 discusses the implications of all these results. In this presentation, I indicate only the important ideas referring to Bhattacharya [8–10] for details.

Before we begin, a word about the notation and conventions which are used unless otherwise mentioned. Scalars are denoted using Greek letters. Roman letters denote vectors, while Roman capital letters denote tensors. $\mathbb{1}$ is the identity tensor. All vectors with hats, example $\hat{\imath}$, are unit vectors. We are usually in 3 dimensions. Subscripts i, j, k, l are integers that run from 1 to 3, while I, J, K, L run between 1 and some specified number. Superscript k is used to index sequences and runs from 1 to infinity. $SO(3) = \{Q | QQ^T = Q^TQ = \mathbb{1};\ \det Q = +1\}$ is the group of all rotations in three dimensions.

2. Constitutive model. Ericksen and James amongst others [6,7,10,21-28,34–36,55,56,69] have developed a constitutive model based on the continuum point of view for crystalline solids that undergo a structural phase transformation. Both the wedge-like microstructure and self-accommodation that we are interested in studying are observed at the transformation temperature. Therefore, I describe a simplified version of the constitutive model where we assume that the temperature is fixed at the transformation temperature.

Consider a specimen of undistorted austenite occupying an open, bounded, connected domain $\Omega \subset \mathbb{R}^3$ with vol $(\partial\Omega) = 0$. Describe all configurations of the crystal as deformations $y : \Omega \to \mathbb{R}^3$ of this reference configuration. For example, the transformation from austenite to martensite is described by the homogeneous deformation $y = U_1 x$ of the reference configuration. The positive-definite symmetric tensor U_1 is called the transformation strain or Bain strain. It depends on the lattice parameters of both the austenite and the martensite and can be measured for any given material.

We want to consider deformations that do not break up the crystal, but at the same time display microstructure. In other words, we want y to be continuous but allow jumps in ∇y $\left(\nabla y \text{ is the tensor whose components in some rectangular coordinate system is given by } (\nabla y)_{ij} = \frac{\partial y_i}{\partial x_j}\right)$. This motivates the assumption that y is Lipschitz or $y \in W^{1,\infty}(\Omega; \mathbb{R}^3)$. Finally, in order to prevent matter from penetrating itself, we assume that y is invertible and that $\det \nabla y > 0$.

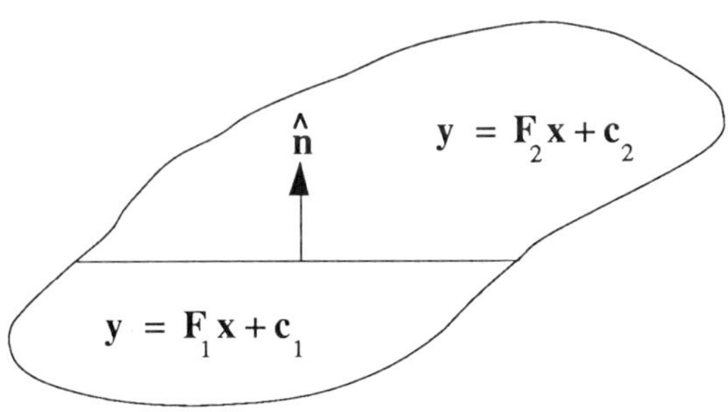

Figure 2.1: The Hadamard Jump Condition

A consequence of the continuity of y is that ∇y cannot jump arbitrarily. Consider for example the piece-wise affine deformation shown in Figure 2.1 where F_1 and F_2 are constant tensors and c_1 and c_2 are suitably chosen constant vectors. Such a deformation y is continuous if and only if

$$F_1 - F_2 = a \otimes \widehat{n} \quad \text{where} \quad (a \otimes \widehat{n})_{ij} = a_i \widehat{n}_j \tag{2.1}$$

for some vectors a and $\widehat{n}$ [11]. This is the well-known Hadamard Jump Condition or Rank-one Compatibility Condition. The interface is necessarily a plane with normal $\widehat{n}$.

Assume that there is a Helmholtz free energy per unit reference volume, ϕ, which depends on the local change in shape measured by the deformation gradient. The total energy of the crystal at configuration y is given by

$$\xi_\Omega[y] = \int_\Omega \phi(\nabla y(x))dx$$

Assume that $\phi(F)$ is defined and is continuous on $\mathcal{D} \subset \mathbb{R}^{3\times 3}$, an open, bounded neighborhood of the identity $\mathbb{1}$. Assume that $U_1 \in \mathcal{D}$ and that $\det F > 0$ on $\mathcal{D}$. $\mathcal{D}$ is a "Pitteri neighborhood" of $\mathbb{1}$.

We do not expect rigid body rotations of the crystal to change its energy. Similarly, the austenite lattice contains crystallographically indistinguishable directions. In order that our free energy reflects these, we make the following assumptions. Firstly, assume that $\mathcal{D}$ satisfies the following invariance.

$$F \in \mathcal{D} \Longrightarrow QFR \in \mathcal{D} \qquad \forall\, Q \in SO(3) \quad \text{and} \quad R \in P_a\,.$$

Here, P_a is the point group of the austenite lattice. The point group of a lattice is the group of rotations that map a lattice back into itself. Corresponding to the different crystal systems, there are seven point groups, all of which are finite. For example, the point group of the cubic lattice consists of the 24 rotations that map a cube back into itself. Now, we assume that $\phi(F)$ has the following invariance.

$$\begin{array}{lll} \text{Frame indifference}: & \phi(F) = \phi(QF) & \forall \quad Q \in SO(3) \\ \text{Material symmetry}: & \phi(F) = \phi(FR) & \forall \quad R \in P_a\,. \end{array} \tag{2.2}$$

Frame indifference ensures that rigid body rotations do not change the energy, while material symmetry reflects the fact that the austenite lattice contains crystallographically indistinguishable directions. The choice of P_a as the material symmetry group is motivated by the molecular theory [22, 55, 56].

As was mentioned earlier, the austenite has greater symmetry than the martensite. Therefore, $RU_1R^T \neq U_1$ for all $R \in P_a$. Define $U_1, U_2, \ldots, U_\nu$ to be the distinct tensors of the form RU_1R^T for $R \in P_a$. The configurations described by the deformation $y = U_K x$, $K = 1, 2, \ldots \nu$ correspond to the different variants of martensite. A consequence of material symmetry is that all variants have the same energy.

Above the transformation temperature, the austenite is stable. Below it, the martensite is stable. At the transformation temperature, there is an exchange of stability between the two phases and we are led to assume that the free energy is minimized at both $\mathbb{1}$ and U_1. By adding a constant if necessary, we assume that the minimum value of ϕ is 0. By imposing frame indifference and material symmetry, we conclude that ϕ is minimized on the sets $\mathcal{A}$ and $\mathcal{M}$ where

$$\begin{array}{ll} \text{Austenite well,} & \mathcal{A} = SO(3) \quad \text{and} \\ \text{Martensite wells,} & \mathcal{M} = \displaystyle\bigcup_{K=1}^{\nu} SO(3)U_K\,. \end{array} \tag{2.3}$$

Finally assume that $\mathcal{A} \cup \mathcal{M}$ are the only minimizers of ϕ. Therefore,

$$\phi(F) \begin{cases} = 0 & F \in \mathcal{A} \cup \mathcal{M} \\ > 0 & F \in \mathcal{D} \backslash (\mathcal{A} \cup \mathcal{M}) \end{cases} \tag{2.4}$$

Now, if we prescribe the displacement boundary condition $y(x) = y_0(x)$ for $x \in \partial\Omega$, the observed configuration is the one that minimizes $\xi_\Omega[y]$ amongst all allowable deformations $y \in \mathcal{S}(y_0)$ where

$$\mathcal{S}(y_0) \overset{\text{def}}{=} \{y \in W^{1,\infty}(\Omega;\mathbb{R}^3) | \nabla y \in \overline{\mathcal{D}} \quad \text{and} \quad y = y_0 \quad \text{on} \quad \partial\Omega\}$$

The assumptions on the free energy density ϕ imply that it is not quasiconvex [18, 36]. Therefore, it is possible to find boundary conditions y_0 for which $\xi_\Omega[y]$ has no minimizer in $\mathcal{S}(y_0)$. This happens even for "acceptable boundary conditions", those for which $\mathcal{S}(y_0)$ is not empty. In such cases, it is possible to find a sequence of deformations $y^k \in \mathcal{S}(y_0)$ for which $\xi_\Omega[y]$ becomes smaller and smaller, going to the infimum in the limit. These minimizing sequences often involve microstructural details on a finer and finer scale and may not converge in the strong sense. However, $\mathcal{D}$ is bounded. Hence, y^k is uniformly bounded in $W^{1,\infty}(\Omega;\mathbb{R}^3)$ and contains at least a subsequence, also called y^k, which converges weak $*$ in $W^{1,\infty}(\Omega;\mathbb{R}^3)$. This is denoted by $y^k \overset{*}{\rightharpoonup} y$ in $W^{1,\infty}(\Omega;\mathbb{R}^3)$ and means

$$\text{(2.5)} \qquad \begin{aligned} &\operatorname*{ess\,sup}_{\Omega} \{|y^k| + |\nabla y^k|\} < M \qquad \forall \;\; k\,, \\ &y^k \longrightarrow y \qquad \text{uniformly in} \quad \Omega \qquad \text{and} \\ &\int_{\Omega'} \nabla y^k(x)dx \longrightarrow \int_{\Omega'} \nabla y(x)dx \qquad \forall \;\; \text{measurable} \quad \Omega' \subset \Omega\,. \end{aligned}$$

It is easy to show that $\mathcal{S}(y_0)$ is weak $*$ closed in $W^{1,\infty}(\Omega;\mathbb{R}^3)$ so that for any $y^k \overset{*}{\rightharpoonup} y$ in $W^{1,\infty}(\Omega;\mathbb{R}^3)$, $\xi_\Omega[y]$ is well-defined. Therefore, the non-attainment means that $\xi_\Omega[y] > \liminf_{k\to\infty} \xi_\Omega[y^k]$ for a minimizing sequence y^k. The fundamental hypothesis of Ball and James [6, 7] is that these weak $*$ convergent minimizing sequences describe the fine scale microstructure that we observe in martensites. In other words, crystals develop fine scale microstructure in an attempt to minimize the energy while satisfying the prescribed boundary conditions.

From (2.5), it is clear that the deformations of a weak $*$ convergent sequence converge uniformly. However, the gradients may develop oscillations and may not converge even pointwise; however, they remain bounded and converge in an average sense. The free energy depends on the deformation gradient. Therefore, in order to study problems involving microstructures, it is necessary to keep track of the gradients. The Young measure serves as exactly such an "accounting device" [5]. We refer to Ball [5] for a suitable version of the Young Measure theorem. For $y^k \overset{*}{\rightharpoonup} y$ in $W^{1,\infty}(\Omega;\mathbb{R}^3)$, the Young measure ν_x gives the probability distribution of the gradients in the neighborhood of the point x.

It is easy to show that a microstructure has zero energy if and only if the Young measure ν_x is supported on the set $\mathcal{A} \cup \mathcal{M}$ a.e. $x \in \Omega$[36] . The Young measure being supported on $\mathcal{A} \cup \mathcal{M}$ a.e. means that the gradients "essentially lie" in $\mathcal{A} \cup \mathcal{M}$ except in a set, the volume of which goes down to zero in the limit. Recall that the transformation strain U_1 determines the sets $\mathcal{A}$ and $\mathcal{M}$. Thus, U_1 determines what kind of microstructure a given material can form.

At this workshop Dr. J.M. Ball discussed at length the microstructures whose Young measure is supported on the set $\mathcal{M}$ which contains two variants. Drs. N.B. Firoozye, V. Sverak and P. Pedregal discussed similar variational problems and analytic methods of studying them while Drs. C. Collins and M. Chipot presented the results of some numerical computations.

3. The wedge

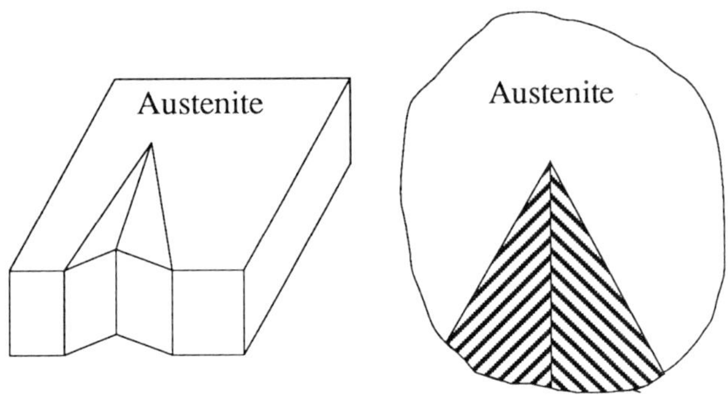

Figure 3.1: A schematic picture of the wedge-like microstructure.

It is very common to observe wedge-like or spear-like microstructure in materials undergoing a reversible martensitic transformation, especially in shape-memory alloys [1, 2, 16, 48–54, 63, 64]. As the material is cooled from above to below the transformation temperature, wedge-shaped regions of martensite grow into the austenite. As shown in Figure 3.1, the wedge consists of two sets of fine martensitic twins (fine laminate consisting of alternating bands of two variants of martensite) separated by a mid-rib. This microstructure provides a simple mechanism for the easy initiation of transformation [53] and is thus important for thermoelasticity and reversibility of transformation. Can every material undergoing a martensitic transformation form a coherent wedge-like microstructure? In other words, is it possible to construct a weak $*$ convergent sequence of deformations, each resembling the wedge, such that the Young measure is supported on $\mathcal{A} \cup \mathcal{M}$ almost everywhere?

It is clear from Figure 3.1 that the wedge-like microstructure contains an interface between a region of martensite and a region of austenite. In order to form an interface between homogeneous austenite and homogeneous martensite, it is necessary according to (2.1) that $A - \mathbb{1} = a \otimes \hat{n}$ for some $A \in \mathcal{M}$. (Recall that we chose austenite as our reference so that $\mathbb{1} \in \mathcal{A}$). For this to be true, U_1 must have an eigenvalue equal to one [6]. This is a very strict condition and does not hold in general. In fact, I know of only one alloy, $Ti - 29$ at. % Ta where Bywater and Christian [14] modified the composition until the middle eigenvalue became equal to one. Therefore, we need more general ways of constructing an austenite–martensite interface.

Consider tensors A and B such that $B - A = a \otimes \widehat{n}$ for some vectors a and $\widehat{n}$. For any $\lambda \in [0,1]$, let $\chi(t) : (0,1) \to [0,1]$ be the characteristic function of $(0, \lambda)$. Extend χ periodically with period 1 to $\mathbb{R}$. Then, for any positive integer k, the deformation

$$y^k(x) = Ax + \int_0^{kx\cdot\widehat{n}} \chi(\tau) d\tau a$$

is as shown in Figure 3.2a. $y^k \in W^{1,\infty}(\Omega; \mathbb{R}^3)$. It is piecewise affine with gradients A and B in the alternating bands. As $k \to \infty$, $y^k \overset{*}{\rightharpoonup} y$ in $W^{1,\infty}(\Omega; \mathbb{R}^3)$ where $y(x) = (\lambda B + (1-\lambda)A)x$. If A and B are in the energy wells, $\xi_\Omega[y^k] = 0 \quad \forall\, k$. Thus y^k is a zero-energy approximation of the homogeneous deformation $(\lambda B + (1-\lambda)A)x$.

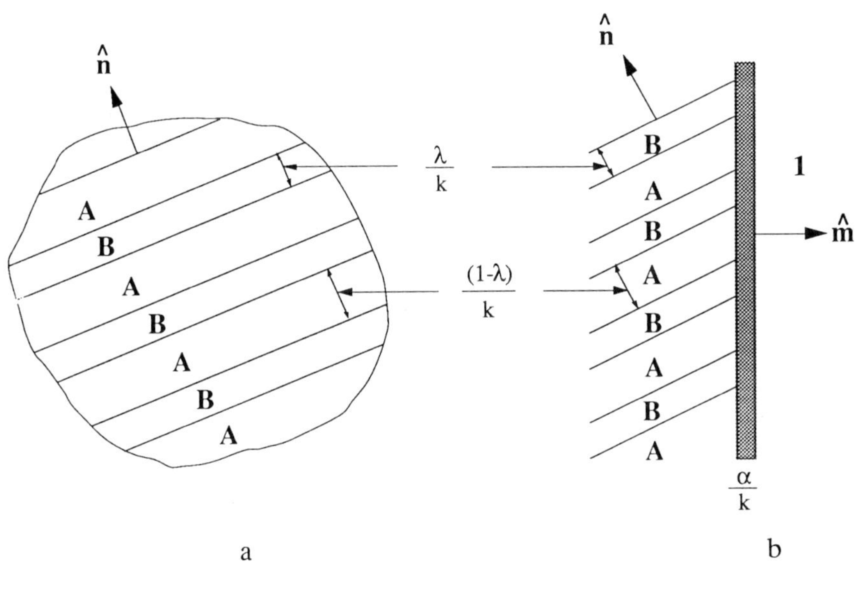

Figure 3.2: The kth term of different sequence of deformations. Figure 3.2 a shows fine twins while Figure 3.2 b shows an approximate interface.

Therefore, even if neither A nor B is compatible with $\mathbb{1}$, it is possible to construct a sequence of deformations as shown in Figure 3.2b which has an interface with alternating bands of two martensitic variants on one side and $\mathbb{1}$ on the other, provided $\lambda B + (1-\lambda)A$ is compatible with $\mathbb{1}$ for some λ. Since neither A nor B is compatible with $\mathbb{1}$, this deformation is not continuous unless an interpolation function is introduced in the shaded region Ω_k. Unfortunately, the gradients of the interpolation function need not lie in the energy wells and the total energy has a non-zero contribution from the interpolation region. Therefore,

$$\xi_\Omega[y^k] = \int_\Omega \phi(\nabla y^k(x))dx = \int_{\Omega_k} \phi(\nabla y^k(x))dx \quad > 0 \,.$$

However, the mismatch decreases as k increases and it is possible to make the interpolation region smaller and smaller while keeping the gradients of the interpolation region uniformly bounded. Thus, $\mathrm{vol}(\Omega_k) \to 0$ while $\phi(\nabla y^k(x))$ remains bounded uniformly in k. Hence $\xi_\Omega[y^k] \to 0$ as $k \to \infty$. Ball and James [6] have shown that the sequence shown in Figure 3.2b is minimizing if and only if $A, B \in \mathcal{M}$ and

$$B - A = a \otimes \widehat{n} \qquad \text{and}$$

$$(\lambda B + (1-\lambda)A) - \mathbb{1} = b \otimes \widehat{m}$$

for some vectors $a, b, \widehat{n}$ and $\widehat{m}$ and for some scalar $\lambda \in [0,1]$. $\widehat{m}$ is the normal to the habit plane or the interface between the fine twins of martensite and the austenite. Thus, in order to construct an approximate interface, it is necessary and sufficient that the average deformation gradients satisfy the Hadamard Jump Condition.

Thus, Figure 3.2b is a way of forming an austenite-martensite interface. Indeed, this is exactly the crystallographic or phenomenological theory of martensite [13, 67]. However, the crystallographic theory is a purely kinematic or geometric theory. The alternating bands of A and B correspond to the lattice invariant shear of the crystallographic theory.

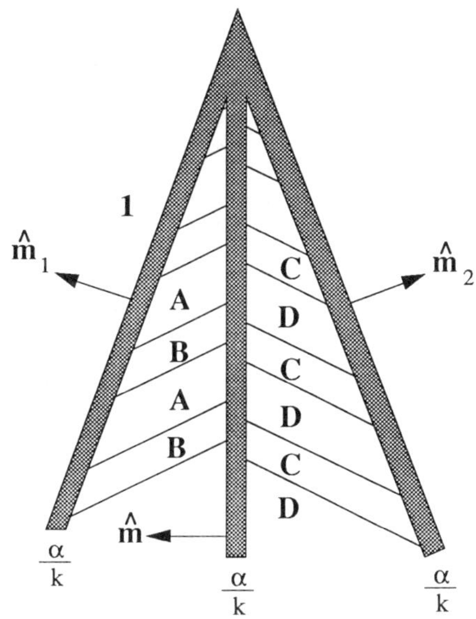

Figure 3.3: A wedge-like sequence of deformations

Consider the wedge-like sequence of deformations shown in Figure 3.3. If A, B, C and D belong to the martensite well $\mathcal{M}$, this sequence is minimizing if and only if

$$\text{(i)} \quad B - A = a_1 \otimes \widehat{n}_1$$

$$\text{(ii)} \quad D - C = a_2 \otimes \widehat{n}_2$$

(3.1) $$\text{(iii)} \quad (\lambda_1 B + (1-\lambda_1)A) = \mathbb{1} + b_1 \otimes \widehat{m}_1$$

$$\text{(iv)} \quad (\lambda_2 D + (1-\lambda_2)C) = \mathbb{1} + b_2 \otimes \widehat{m}_2$$

$$\text{(v)} \quad (\lambda_2 D + (1-\lambda_2)C) - (\lambda_1 B + (1-\lambda_1)A) = b \otimes \widehat{m}$$

for some $\lambda_1, \lambda_2 \in [0,1]$ and vectors $a_1, a_2, b_1, b_2, b, \widehat{n}_1, \widehat{n}_2, \widehat{m}_1, \widehat{m}_2$ and $\widehat{m}$. Substituting (3.1) (iii) and (3.1) (iv) in (3.1) (v) gives

$$b_1 \otimes \widehat{m}_1 - b_2 \otimes \widehat{m}_2 = b \otimes \widehat{m} \ .$$

This is possible if and only if b_1 is parallel to b_2 or $\widehat{m}_1$ is parallel to $\widehat{m}_2$. However, in order to maintain the wedge-like geometry, $\widehat{m}_1$ and $\widehat{m}_2$ can not be parallel. Thus, a material can form a wedge if and only if (3.1) (i)–(iv) hold and b_1 is parallel to b_2[8].

Recall that A, B, C and D belong to the martensite wells $\mathcal{M}$. Therefore, each of them is of the form QU_K for $Q \in SO(3)$ and $K \in \{1, 2, \ldots, \nu\}$. It turns out that under rather minor restrictions on U_1, (3.1) (i)–(iv) can be solved for all unknowns in terms of the transformation strain U_1[6]. In particular, b_1 and b_2 are obtained as functions of U_1. The fact that two vector valued functions of U_1 have to be parallel to each other imposes strict restrictions on U_1. Thus, only very special materials, those whose transformation strain satisfy these restrictions, can form a wedge. For example, in a cubic to tetragonal transformation,

$$U_1 = \begin{pmatrix} \eta_2 & 0 & 0 \\ 0 & \eta_1 & 0 \\ 0 & 0 & \eta_1 \end{pmatrix} = \text{Diag}\,[\eta_2, \eta_1, \eta_1]$$

in an orthonormal basis parallel to the edges of the cubic unit cell. Above $\eta_2 > 0$ and $\eta_1 > 0$ may be determined experimentally for any given material. It is possible to solve (3.1) (i)–(iv) and form austenite-martensite interfaces as in Figure 3.2b if and only if [6]

$$\eta_1 < 1 < \eta_2 \qquad \frac{1}{\eta_1^2} + \frac{1}{\eta_2^2} \leq 2$$

or

$$\eta_2 < 1 < \eta_1 \qquad \eta_1^2 + \eta_2^2 \leq 2\,.$$

The material can form a wedge if and only if [8]

$$\eta_1^2 = \frac{(1-\eta_2^2)^2 + 4\eta_2^2(1+\eta_2^2)}{(1-\eta_2^2)^2 + 8\eta_2^4} \tag{3.2}$$

These relations are plotted in Figure 3.4. In the $\eta_1 - \eta_2$ plane, the material can form a wedge if and only if (η_1, η_2) lies on the solid line.

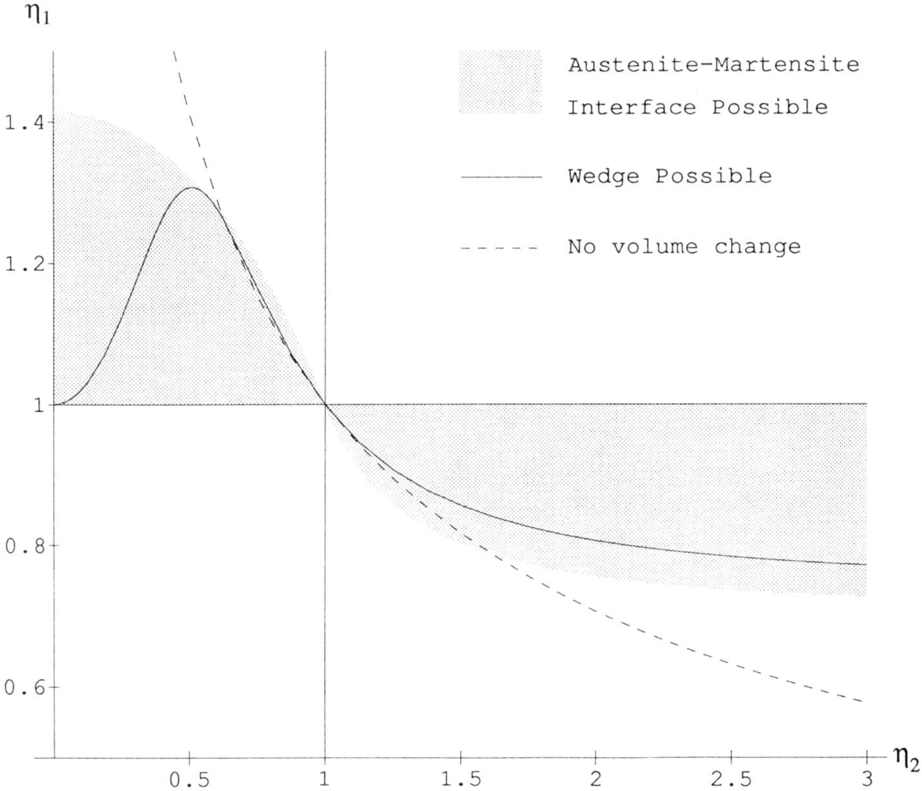

Figure 3.4: Special relation for the wedge in the cubic to tetragonal case. If the transformation strain of the material lies in the shaded region, the material can form an austenite-martensite interface. If it lies on the solid curve, it can form a wedge. If it lies on the dashed curve, there is no volume change during transformation. The two curves intersect at the points (1,1) and (0.6272,1.2627).

Similarly, in a cubic to orthorhombic transformation,

$$U_1 = \begin{pmatrix} \dfrac{\alpha+\gamma}{2} & 0 & \dfrac{\alpha-\gamma}{2} \\ 0 & \beta & 0 \\ \dfrac{\alpha-\gamma}{2} & 0 & \dfrac{\alpha+\gamma}{2} \end{pmatrix}$$

in an orthonormal basis parallel to the edges of the cubic unit cell. Above, $\alpha > 0$, $\beta > 0$, $\gamma > 0$ are experimentally determinable for any given material. In this case, it is possible for a material to form a wedge if and only if it lies on a small family of surfaces in $\alpha - \beta - \gamma$ space [8].

A comment is in order. It should be reiterated that the conditions (3.1) are necessary only for the special microstructure assumed in Figure 3.3. It assumes that the wedge-like microstructure involves four tensors from the martensite wells $\mathcal{M}$ and $\mathbb{1}$ from the austenite well $\mathcal{A}$. Moreover, it assumes that A and B as well as C and D are finely twinned and each set of fine twins form an interface with $\mathbb{1}$. However, there is no a priori assumption about the proportions of the fine twins or the normals to the interfaces. They come out as parts of the solution of (3.1) (i)–(iv). All but one experimental investigation of the wedge that I found in the literature [1, 2, 16, 48–54, 64] clearly show the geometry assumed here. The only exception is the recent observation of Tan and Xu [63] of the wedge-like microstructure in Cu–14 wt. % Al - 4.2 wt. % Ni. Professor I. Müller showed slides of their observations during the workshop. One of their optical micrographs suggests that the fine twins themselves have some substructure. The nature of the substructure is not clear from the micrograph. Moreover, Professor Xu notes [68] that "this kind of internal twins was a freak observation and appeared only in some locations of very large martensite arrows".

4. The shape-memory effect and self-accommodation. Shape-memory effect is the ability of a material to remember its shape above a critical temperature. Below the critical temperature, the alloy is extremely malleable – undergoing apparently plastic deformations with strains as large as 10% under very small forces. However, all the strain is recovered when it is heated to above the critical temperature. Cooling from above to below the critical temperature does not cause any macroscopic shape change and the whole cycle can be repeated. This is shown schematically in Figure 4.1.

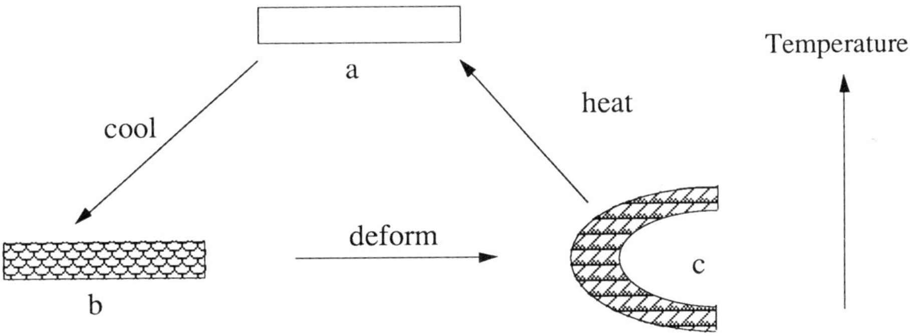

Figure 4.1: The shape-memory effect

The shape-memory effect is observed in some materials that undergo the martensitic phase transformation. The critical temperature is the transformation temperature. Consider a specimen of a given shape in Figure 4.1 a. It is in the austenite phase. On cooling, the austenite transforms to the martensite. However, the variants of martensite arrange themselves in such a way that there is no macroscopic change in shape (Figure 4.1 b). Thus, though at the microscopic level there is a change in shape due to the structural transformation, the different variants "accommodate each others strain" and there is no observable change in shape at the macroscopic level. This is known as "self-accommodation". Below the transformation temperature, the martensite can deform by converting one variant into another and forming a new microstructure (Figure 4.1 c). On heating, every variant converts back to the austenite. Since there is only one variant of austenite, all the strain is recovered and the specimen goes back to its original shape.

In this section, we study self-accommodation. By a self-accommodating microstructure, we mean a microstructure of martensite which satisfies identity boundary condition. Recall that we chose austenite as our reference configuration. To be precise, a self-accommodating microstructure is a sequence of deformations y^k which has the following properties

(i) $y^k \overset{*}{\rightharpoonup} y \quad$ in $\quad W^{1,\infty}(\Omega; \mathbb{R}^3)$

(ii) $y^k(x) = x \qquad x \in \partial\Omega$

(iii) $\operatorname{supp} \nu_x \subset \mathcal{M}$ a.e. $x \in \Omega$.

Here, ν_x is the Young measure delivered by the fundamental theorem on Young measures (Ball [5]).

Apart from the fact that self-accommodation is an inherent part of the shape-memory phenomenon, it may also be important for the reversibility of transformation in polycrystals and for the easy nucleation of martensite during cooling. A polycrystal is made up of numerous randomly oriented crystals or grains. If the grains change their shape during transformation, owing to the random arrangement it is very likely that there will be some mismatch at the grain boundaries resulting in internal stresses and cracks. Such a transformation is unlikely to be reversible. The possible presence of temperature gradients during transformation worsens the situation. On the contrary, if the material is self-accommodating, there would be no displacement and hence no mismatch at the grain boundaries. Such a transformation will be easily reversible. Similarly, during cooling, if the nucleus of martensite forms in the interior of the crystal, it is an island of martensite surrounded completely by the austenite. If this microstructure is incoherent or introduces large internal stresses, such internal nucleation is unlikely. Thus, internal nucleation is likely to occur only in self-accommodating materials.

Studying common shape-memory alloys, Tas, Delaey and Deruyterre [62], Saburi and Wayman [57] and Miyazaki, Otsuka and Wayman [43] proposed certain microstructure as self-accommodating. The central idea of all their analysis is that of a "self-accommodating plate group." It turns out that these microstructure violate certain jump or compatibility conditions [9]. Moreover, their analysis does not determine which material can and which material can not form a self-accommodating microstructure. It is clear that not every material undergoing a martensitic transformation can do so. For example, if the volume of the martensite is smaller than that of the austenite, it will not be possible to surround an unbroken, unstressed domain of martensite by a sea of unstressed austenite. Such a material is not self-accommodating. Which are the materials that can form a self-accommodating microstructure?

Table 4.1 lists the necessary and sufficient conditions that the lattice parameters (transformation strain) of a material have to satisfy in order that the material can form a self-accommodating microstructure [9]. Quite interestingly, even though it is the martensite that forms the self-accommodating microstructure, the conditions as stated in Table 4.1 depend on the symmetry of the austenite and not that of the martensite

Table 4.1: Necessary and sufficient conditions for self-accommodation

Symmetry of the austenite	Necessary and Sufficient conditions for self-accommodation
Cubic	$\det U_1 = 1$
Tetragonal	(i) $\det U_1 = 1$ and (ii) $\dfrac{1}{E_{22}E_{33} - E_{23}^2} \leq 1 \leq E_{11}$
Orthorhombic	(i) $\det U_1 = 1$, (ii) $\dfrac{1}{E_{22}E_{33} - E_{23}^2} \leq 1 \leq E_{11}$, (iii) $\dfrac{1}{E_{11}E_{33} - E_{13}^2} \leq 1 \leq E_{22}$ and (iv) $\dfrac{1}{E_{11}E_{22} - E_{12}^2} \leq 1 \leq E_{33}$.
Monoclinic	The two well problem [7]

U_1 is the transformation and $E = U_1^2$. The components of E are expressed in an orthonormal basis parallel to the cubic/tetragonal/orthorhombic unit cell of the austenite. In the case of the tetragonal austenite, $\widehat{\mathbf{i}}_1$ is the "c-axis" of the tetragonal lattice.

As explained above, volume preservation during transformation is necessary for self-accommodation. According to Table 4.1, volume preservation during transformation is also sufficient if the symmetry of the austenite is cubic. In fact, in the case of cubic austenite, it is always possible to construct a microstructure which is a pure dilatation with respect to the austenite. The amount of dilatation of course depends on the transformation volume change. However, if the symmetry of the austenite is not cubic, the lattice parameters of the material have to satisfy additional conditions which are extremely stringent and non-generic. For example, in a tetragonal to orthorhombic transformation,

$$U_1 = \text{Diag } [\alpha, \beta, \gamma] \tag{4.1}$$

in an orthonormal basis $\{\widehat{i}, \widehat{i}_2, \widehat{i}_3\}$ parallel to the edges of the tetragonal unit cell. If

$\widehat{\imath}_1$ is the c-axis of the tetragonal lattice, for self-accommodation it is necessary and sufficient that

$$\alpha = \beta\gamma = 1 \tag{4.2}$$

Similarly, in an orthorhombic to monoclinic transformation

$$U_1 = \begin{pmatrix} \alpha & 0 & 0 \\ 0 & \beta & \delta \\ 0 & \delta & \gamma \end{pmatrix}$$

in an orthonormal basis $\{\widehat{\imath}_1, \widehat{\imath}_2, \widehat{\imath}_3\}$ parallel to the edges of the orthorhombic unit cell. $\widehat{\imath}_1$ is the axis of monoclinic symmetry. According to Table 4.1, the material can form a self-accommodating microstructure if and only if

$$\alpha = 1$$

$$\beta\gamma - \delta^2 = 1$$

$$\beta^2 + \delta^2 \geq 1$$

$$\gamma^2 + \delta^2 \geq 1$$

It should be emphasized that results in Table 4.1 are derived with no a priori assumption on the microstructure. Therefore, unless they are satisfied, no matter how complicated the microstructure is, it will not be self-accommodating.

The proof of these results can be found in Bhattacharya [9]. The necessity is a consequence of the "minors relations". These relations are just the statement of the fact that the sub-determinants or minors of a matrix are null–lagrangeans and sequentially weak* continuous. The sufficiency is demonstrated by constructing a self-accommodating microstructure which consists of multiple levels of laminates.

5. Comparison of theory and experiment. In order to compare theory and experiment, it is necessary to calculate the transformation strain from the lattice parameters of both the austenite and the martensite measured at the transformation temperature. Unfortunately, such measurements are not always found in the literature. Often, all that is found are parameters for the austenite measured by one researcher at some temperature and those for the martensite measured by a different researcher at a different temperature. Similarly, shape-memory is observed by one researcher while the lattice parameters are measured by others. Strictly speaking, the results in §3 and §4 are true only at the transformation temperature. The lattice parameters change with changes in temperature. Similarly, they change with changes in composition, which unavoidably differs in specimens used by different researchers. One should keep this in mind while making any comparisons.

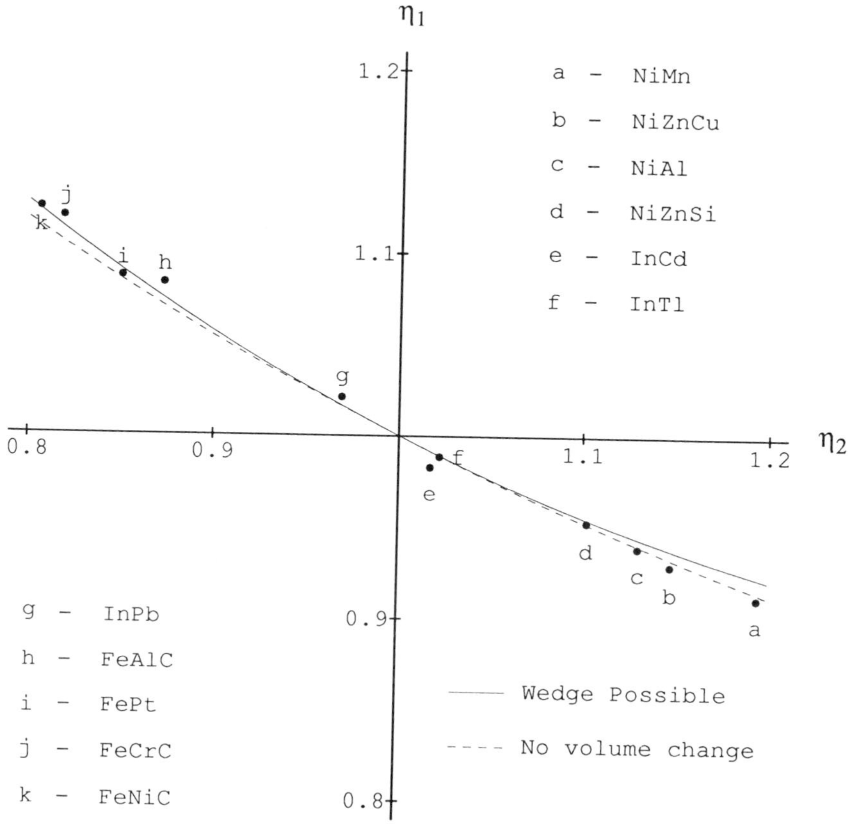

Figure 5.1: Comparison of theory and experiment for alloys undergoing cubic to tetragonal transformation

Figure 5.1 shows the $\eta_1 - \eta_2$ space of materials that undergo cubic to tetragonal transformation. The solid line is the line on which it is possible to form a wedge, i.e., η_1 and η_2 satisfy (3.2). The dashed line is the line on which there is no transformation volume change (i.e., $\eta_1^2\eta_2 = 1$) as required for self-accommodation. The various points are the measured lattice parameters of actual alloys. The details are explained in Table 5.1. Substituting the measured value of η_2 in (3.2), we obtain a theoretical value of η_1 for forming the wedge. This is compared with the measured value of η_1 in Table 5.1. This table also lists the percentage volume change $(100(1-\det U_1))$ during the transformation from austenite to martensite. Table 5.2 compares the theory and experiment for alloys undergoing cubic to orthorhombic transformation. Corresponding to the measured value of α and γ, a theoretical value of β for forming a wedge is read off from one amongst the family of surfaces on which it is possible to form a wedge. This is compared to the measured value of β in Table 5.2. This table also lists the percentage volume change during transformation. Finally, Table 5.3 lists the percentage volume change in some common shape-memory alloys that undergo the cubic to monoclinic transformation.

Table 5.1: Comparison of theory and experiment for cubic to tetragonal materials

Alloys	Experimental Lattice Parameters η_2	η_1	η_1 for wedge	Percentage volume change	Comments and References
Ni - 36% A1	1.1302	0.9392	0.9445	0.305	Wedges seen Shape-memory [15,16,20]
Ni - 49.4% Mn	1.1941	0.9118	0.9232	0.725	Wedges seen [1,2,39]
Ni - 48% Zn -20%Cu	1.1477	0.9296	0.9383	0.820	Wedges seen Shape-memory Lattice parameters were read of a graph. [45,46]
Ni - 34% Zn -15% Si	1.1026	0.9530	0.9548.	-0.139	Wedges seen? Shape-memory? Lattice parameters were read off a graph. [46]
In - 22.73% T1	1.0221	0.9889	0.9898	0.046	Shape-memory [31,47]
In - 32% Pb	0.9688	1.0208	1.0162		Wedges seen Shape-memory Lattice parameters were read off a graph. [37,47]
In - 5.6% Cd	1.0170	0.9830	0.9177		Wedges seen Shape-memory Lattice parameters were read off a graph. [38,47]
Fe - 31% Ni -0.3% C	0.8059	1.1241	1.1227	-1.833	Wedges seen [48]
Fe - 7.9% Cr - 1.1% C	0.8184	1.1195	1.1132	-2.568	[44]
Fe - 7% A1 -2% C	0.8727	1.0833	1.04745	-2.24	[66]
Fe - 24% Pt	0.8441	1.0870	1.0944	0.263	Shape-memory [29,61,64]
Fe - 31.2% Pd				0.2	Shape-memory [59]

All compositions are in atomic percentages unless otherwise mentioned.
Percentage volume change $= 100(1 - \det \mathbf{U}_1)$.

Table 5.2: Comparison of theory and experiment for cubic to orthorhombic materials

Alloys	Experimental Lattice Parameters α, γ	β	β for wedge	Percentage volume change	Comments and References
Cu - 14.2 wt % A1 - 4.3 wt % Ni	$\alpha = 1.0619$ $\gamma = 1.0230$	0.9178	0.9193	0.297	Shape-memory. Wedges seen very often seen. [19,50-54,63]
Au - 47.5 % Cd	$\alpha = 1.0138$ $\gamma = 1.0350$	0.9491	0.9508	0.413	Shape-memory. [12,17]
Ag - 52 wt % Cd	$\alpha = 1.0427$ $\gamma = 1.0191$	0.9396	0.9525	0.157	Shape-memory.[41]
Zr - 19.5 % Th	$\alpha = 1.069$ $\gamma = 1.014$	0.902	0.9487	2.22	[32]

All compositions are in atomic percentages unless otherwise mentioned.
Percentage volume change = $100(1 - \det \mathbf{U}_1)$.

Table 5.3: Comparison of theory and experiment for cubic to monoclinic materials

Alloys	Percentage volume change	Comments and References
Ni - 50 % Ti	0.023	Most common shape memory alloy.[40,43]
Cu - 17.2% A1 - 13.3 % Zn	0.356	Shape memory[57]
Cu - Zn -Ga	1.06	Shape memory[57]
Cu - A1	"approx. zero"	Shape memory[62]

All compositions are in atomic percentages unless otherwise mentioned.
Percentage volume change = $100(1 - \det \mathbf{U}_1)$.

It is clear that common shape-memory alloys like $Ni-Ti$, $Cu-A\ell-Ni$, $Ni-A\ell$, $Cu-A\ell$ and $Cu-Zn-A\ell$ have a cubic austenite and small transformation volume change. Moreover, the volume seems to decrease during the transformation from austenite to martensite. Often, the lattice parameters of the austenite are measured at a higher temperature while those of the martensite are measured at a lower temperature. Therefore, part of the volume change may be attributed to thermal expansion. Ferrous martensites which are not shape-memory alloys have comparatively larger volume change. Moreover, the volume increases as we go from austenite to martensite. Zirconia undergoes a tetragonal to monoclinic transformation accompanied by a volume change of approximately 3 % [60]. This transformation often results in the cracking of the specimen. The observations of Arlt [4] in polycrystals of $BaTiO_3$ and $YBa_2Cu_3O_{7-\delta}$ are worth mentioning. $BaTiO_3$ undergoes a volume preserving cubic to tetragonal martensitic transformation [30]. $YBa_2Cu_3O_{7-\delta}$ undergoes a first order tetragonal to orthorhombic transformation which may be induced by both temperature and diffusion of oxygen [3]. Arlt takes aged polycrystals in which the transformation from the high to low symmetry phase is complete. He carefully cuts out a slice in order to observe the microstructure deep in the interior of the polycrystal. He studies the microstructure using both optical and electron microscopes. The grains of $BaTiO_3$ show no cracks and there is no mismatch at the boundaries indicating self-accommodation. In contrast, there is significant cracking and grain boundary mismatch in $YBa_2Cu_3O_{7-\delta}$. The measured U_1 = Diag $[0.9887, 0.9898, 1.0068]$ [3] clearly does not satisfy the necessary and sufficient conditions for self-accommodation (4.2). Arlt also remarks that the cracks lie in a plane perpendicular to $\widehat{\imath}_1$ indicating that they arise as a result of the shrinkage in the direction of the c-axis. These observations in $YBa_2Cu_3O_{7-\delta}$ must be tempered with the remarks of Van Tendeloo [65] who believes that cracking in $YBa_2Cu_3O_{7-\delta}$ is caused by the diffusion of oxygen in the presence of moisture. Thus, we find that in general the experimental observations agree quite well with the results of §4.

Similarly, the lattice parameters of the materials in which wedges are seen satisfy the special relationship for the wedge rather well. Wedges are commonly seen and extensively investigated in $Cu - A\ell - Ni$ which transforms from a cubic DO_3 type austenite to a orthorhombic Cu_3Ti type martensite [19]. As remarked in §3, the normals and other crystallographic details are obtained as solutions of (3.1) (i)–(iv). Table 5.4 shows good agreement between the theoretically predicted and experimentally observed crystallographic details of the microstructure in $Cu - A\ell - Ni$.

Table 5.4: Comparison of theoretically predicted and observed microstructure of Cu-A1-Ni.

	Predicted	Observed
Compound Twins[33]		
Twin Planes	{1,0,0}	{1,0,0}
Twinning shear direction	$\langle -0.018, 0, -0.998\rangle$	$\langle -0.018, 0, -0.9998\rangle$
Twinning shear magnitude	0.0747	0.0744
Austenite-martensite interface	Not possible	Not seen
Type I Twins[33,53]		
Twin Planes	{1,1,0}	{1,1,0}
Twinning shear direction	$\langle -0.736, 0.645, -0.203\rangle$	$\langle -0.736, 0.645, -0.203\rangle$
Twinning shear magnitude	0.261	0.261
Austenite-martensite interface	{0.189, - 0.748, 0.636}	Nearly {3,-3,1}
Direction of the shape strain of the mart.	$\langle 0.248, 0.685, 0.685\rangle$	Not available
Magnitude of the shape strain of the mart.	0.09588	Not available
Midrib	{1,1,0}	{1,1,0}
Type II Twins[33,51]		
Twin Planes	{-0.688, 0.688,-0.228}	{-0.688, 0.688,-0.227}
Twinning shear direction	$\langle 0.75, 0.661, 0.014\rangle$	$\langle 0.75, 0.661, 0.014\rangle$
Twinning shear magnitude	0.261	0.261
Austenite-martensite interface	{0.260, - 0.728, 0.635}	{0.282, -0.741, 0.609}
Direction of the shape strain of the mart.	$\langle 0.130, 0.702, 0.701\rangle$	$\langle 0.214, -0.684, 0.698\rangle$
Magnitude of the shape strain of the mart.	0.09320	0.096
Midrib	{1,1,0}	{1,1,0}
Note: All the indices are refered to the cubic basis. Therefore, in the orthorhombic basis, the normals of the compound twin planes belong to the {110} family and those of the type I twins belong to the {121} family.		

6 Conclusions and discussion. The precise results derived in §3 and §4, along with their good agreement with experimental observations, indicate that the model described in §2 is quite satisfactory in studying crystalline solids that undergo martensitic transformation.

In §3, we show that a commonly observed wedge-like microstructure is possible only in special materials, those whose lattice parameters satisfy certain strict restrictions. This shows that the microstructure may depend very delicately on the lattice parameters. A small change in lattice parameters could result in a drastic change in the microstructure, and thereby macroscopic properties.

§4 discusses the problem of self-accommodation – can a microstructure of martensite be coherently embedded in a sea of austenite? Self-accommodation is an inherent part of the shape-memory effect. It may also be important for the reversibility of transformation in polycrystals and the easy nucleation of martensite during cooling. The main result, Table 4.1, says that a material with cubic austenite is self-accommodating if and only if the transformation volume change is zero. If the symmetry of the austenite is not cubic, the material has to satisfy additional very strict conditions in order to be self-accommodating. Therefore, it seems reasonable to conclude that only materials with cubic austenite which undergo volume preserving transformation can be self-accommodating and hence potential candidates for shape-memory alloys.

Finally, it is interesting to note from Figure 5.1 that in the case of a cubic to tetragonal transformation, the curve on which a wedge-like microstructure is possible and the curve on which the volume is preserved during transformation are quite close when (η_1, η_2) is close to (1,1). Similarly, in the case of a cubic to orthorhombic transformation, large regions of the surface of zero volume change lie close to one amongst the family of surfaces on which a wedge is possible [9]. The lucky accident that both these relations are close to each other means that a self-accommodating material can form a wedge with small elastic energy. Thus it possesses a mechanism for the easy initiation of transformation. Therefore, these materials are ideal candidates for shape-memory alloys. In fact small transformation volume change is one of the empirical criterions currently being used for inventing ferrous shape-memory alloys [42].

Acknowledgement. I thank Dr. Richard D. James for his constant encouragement, numerous suggestions and for comments on an early draft. I am glad to acknowledge the support of the National Science Foundation and the Air Force Office of Scientific Research through NSF/DMS–8718881.

REFERENCES

[1] Adachi, K. and Wayman, C.M., 1985, *Transformation behaviour of nearly stoichiometric Ni-Mn alloys*, Metallurgical Transactions A, 16 A, 1567–1579.

[2] Adachi, K. and Wayman, C.M., 1985, *Electro microscopy study of q-phase martensites in Ni-Mn alloys*, Metallurgical Transactions, A 16 A, 1581–1597.

[3] Andersen, N. H., Lebech, B. and Poulsen, H. F., 1990, *The structural phase diagram and oxygen equilibrium partial pressure of $YBa_2Cu_3O_{6+x}$ studied by neutron powder diffraction and gas volumetry*, Physica C, 172, 31–42.

[4] Arlt, G., 1990, *Twinning in ferroelectric and ferroelastic ceramics: stress relief*, Journal of Materials Science, 22, 2655–2666.

[5] Ball, J.M., 1988, *A version of the fundamental theorem for Young measures, PDEs and continuum models of phase transitions, (ed. Rascle, M., Serre, D. and Slemrod, M.)*, Lecture notes in physics, 344, Springer -Verlag, 207–215.

[6] Ball, J.M. and James, R.D., 1987, *Fine phase mixtures as minimizers of energy*, Archive for Rational Mechanics and Analysis, 100, 15–52.

[7] Ball, J.M. and James, R.D., 1991, *Proposed experimental tests of a theory of fine microstructure and the two well problem*, to appear in Philosophical Transactions of the Royal Society of London, Series A.

[8] BHATTACHARYA, K., 1991, *Wedge-like microstructure in martensite*, to appear in Acta Metallurgica.

[9] BHATTACHARYA, K., 1991, *Self-accommodation in martensites*, IMA Preprint.

[10] BHATTACHARYA, K., 1991, Ph. D Thesis, University of Minnesota, Minneapolis.

[11] BILLINGTON, E. W. AND TATE, A., 1960, *The Physics of Deformation and Flow*, Mc-Graw Hill.

[12] BIRNBAUM, H.K. AND READ, T.A., 1960, *Stress induced twin boundary motion in AuCd β' and β" alloys*, Transactions of the Metallurgical Society of AIME, 218, 662–669.

[13] BOWLES, J.S. AND MACKENZIE, J.K., 1954, *The crystallography of martensitic transformations I and II*, Acta Metallurgica, 2, 129-137, 138–147.

[14] BYWATER, K.A. AND CHRISTIAN, J.W., 1972, *Martensitic transformations in titanium-tantalum alloys*, Philosophical Magazine, 25, 1249–1272.

[15] CHAKRAVORTY, S., 1975, Thesis, University of Illinois, Urbana-Champaign.

[16] CHAKRAVORTY, S. AND WAYMAN, C.M., 1976, *The thermoelastic martensitic transformation in b' Ni-Al alloys: I. Crystallography and morphology and II. Electron microscopy*, Metallurgical Transactions A, 7A , 555–568 and 569–582.

[17] CHANG, L.C. AND READ, T. A., 1951, *Plastic deformation and diffusionless phase changes in metals - the gold-cadbim beta phase*, Journal of metals, Transactions AIME, 191, 47–52.

[18] DACAROGNA, B., 1989, *Direct methods in the calculus of variations*, Springer-Verlag.

[19] DUGGIN, M.J. AND RACHINGER, W.A., 1964, *The nature of the martensitic transformation in a copper-aluminum-nickel alloy*, Acta Metallurgica, 12, 529–535.

[20] ENAMI, K. AND NENNO, S., 1971, *Memory effect in Ni-36.8 at. pct. Al martensite*, Metallurgical Transactions, Vol 2, 1487–1490.

[21] ERICKSEN, J.L., 1977, *Special topics in elastostatics*, Advances in Applied Mechanics, Academic Press 7, 189–243.

[22] ERICKSEN, J.L., 1979, *On the symmetry of deformable crystals*, Archive for Rational Mechanics and Analysis, 72, 1–13.

[23] ERICKSEN, J.L., 1980, *Some phase transitions in crystals*, Archive for Rational Mechanics and Analysis, 73, 99–124.

[24] ERICKSEN, J.L., 1984, *The Cauchy and Born hypotheses for crystals, Phase transformations and material instabilities in solids* (ed. Gurtin, M.), Academic Press, 61–78.

[25] ERICKSEN, J.L., 1985, *Some surface defects in unstressed thermoelastic solids*, Archive for Rational Mechanics and Analysis, 88, 337–345.

[26] ERICKSEN, J.L., 1986, *Stable equilibrium configurations of elastic crystals, Archive for Rational Mechanics and Analysis*, 94, 1–14.

[27] ERICKSEN, J.L., 1989, *Weak martensitic transformations in Bravais lattices*, Archive for Rational Mechanics and Analysis, 107, 23–36.

[28] ERICKSEN, J.L., 1990, *Local bifurcation theory for thermoelastic Bravais latices*, to appear.

[29] FOOS, M., FRANTZ, C. AND GANTIOS, M., 1975, *Shape memory effect and an elasticity associated with the martensitic transformation in the stoichiometric Fe3Pt alloy, Shape Memory Effects in Alloys* (ed. Perkins, J.), Plenum Press.

[30] FORSBERGH, P.W., 1949, *Domain structures and phase transitions in Barium Titanate*, Physical Review, 76 1187–1201.

[31] GUTTMAN, L., 1950, *Crystal structures and transformations in Indium-Thallium solid solutions*, Journal of metals, Transactions AIME, 188, 1472–1477.

[32] HANSON, C.G., RIVLIN, V.G. AND HATT, B.A., 1964, *The β-phase transformation of some zirconium-thorium alloys*, Journal of Nuclear Materials, 12, 83–93.

[33] ICHINOSE, S., FANATSU,Y. AND OTSUKA, K., 1985, *Type II deformation twinning in γ_1 martensite in a Cu-Al-Ni alloy*, Acta Metallurgica, 33, 1613–1620.

[34] JAMES, R.D., 1986, *Displacive phase transformations in solids*, Journal of Mechanics and Physics of Solids, 34, 359–394.

[35] James, R.D., 1987, *The stability and metastability of quartz, Metastability and incompletely posed problems*, IMA, Vol 3, Springer-Verlag, 147–176.

[36] James, R.D. and Kinderlehrer, D., 1989, *PDEs and continuum models of phase transitions,* (ed. Rascle, M., Serre, D. and Slemrod, M.), Lecture notes in physics, 344, Springer -Verlag, 51–84.

[37] Koyoma, Y., Ukena, T. and Nittono, O., 1980, *Phase transformations and shape memory effect in Indium Lead alloys*, Journal of the Japan Institute of Metals, 44, 431–1439.

[38] Koyoma, Y. and Nittono, O., 1979, *Shape memory in In-Cd alloys*, Journal of the Japan Institute of Metals, 43, 262–270.

[39] Krasevec, V., 1975, *The complex microstructure in quenched Ni-Mn alloy*, Physica Status Solidi A, 30, 241–250.

[40] Kudoh, Y., Tokonami, M., Miyazaki, S. and Otsuka, K., 1985, *Crystal structure of the martensite in Ti 49.2 at % Ni alloy analyzed by the single crystal X-ray diffraction method*, Acta Metallurgica, 33, 2049–2056.

[41] Masson, D.B. and Barrett, C.S., 1958, *Effect of deformation and low temperatures on the structure of AgCd and AuZn*, Transactions of the Metallurgical Society of AIME, 212, 260–265.

[42] Miyazaki, S. and Otsuka, K., 1989, *Development of Shape Memory Alloys*, ISIJ International, 29, 353–377.

[43] Miyazaki, S., Otsuka, K. and Wayman, C.M., 1989, *The shape memory mechanism associated with the martensitic transformation in TiNi alloys - I. Self- accommodation and II. Variant Coalescence and shape recovery*, Acta Metallurgica, 37, 1873-1884 and 1885–1890.

[44] Morton, A.J. and Wayman, C.M., 1967, *Theoritical and experimental aspects of the "(225)" austenite-martensite transformation in iron alloys*, Acta Metallurgica, 14, 1567–1581. (47).

[45] Murakami, Y. and Kanchi, S., 1977, *Microscopic observation of thermoelastic α_2 Ni-Zn-Cu martensitic*, Transactions of the Japan Institute of Metals, 18, 423–426.

[46] Murakami, Y., Kanchi, S. and Shimizu, S., 1976, *Martensitic transformation and elastic properties of the ternary Ni-Zn $\beta 2$ base phase alloy, New aspects of martensitic transformations,*, Supplement to the Transactions of the Japan Institute of Metals, 17, 147–157.

[47] Nittono, O. and Koyoma, Y., 1981, *Crystal structure and phase transformations in Indium rich solid solution*, Science reports of the Research Institutes of Tohoku University, A 29, Supplement 1, 53–60.

[48] Okamoto, H., Oka, M. and Tamura, I., 1978, *Coupling of thin-plate martensites in an Fe-Ni-C alloy*, Transactions of the Japan Institute of Metals, 19, 674–684.

[49] Okamoto, K., Ichinose, S., Morii, K., Otsuka, K. and Shimizu, K., 1986, *Crystallography of $\beta 1 - \gamma_1$ stress induced martensitic transformation in a Cu-Al-Ni alloy*, Acta Metallurgica, 34, 2065–2073.

[50] Otsuka, K., 1971, *Origin of memory effect in Cu-Al-Ni alloy*, Japanese Journal of Applied Physics, 10, 571–579.

[51] Otsuka, K., 1986, *Crystallography of martensitic transformation and type II twinning*, Proceedings of the International Conference on Martensitic Transformations (ICOMAT-86), 35–42.

[52] Otsuka, K. and Shimizu, K., 1969, *Morphology and crystallography of thermoelastic g Cu-Al-Ni martensite*, Japanese Journal of Applied Physics, 8, 1196–1204.

[53] Otsuka, K. and Shimizu, K., 1974, *Morphology and crystallography of thermoelastic Cu-Al-Ni martensite analyzed by the phenomenological theory*, Transactions of the Japan Institute of Metals, 15, 103–108.

[54] Otsuka, K. and Shimizu, K., 1975, *Optical and electron microscope observations of transformation and deformation characteristics in Cu-Al-Ni marmem alloys, Shape Memory Effects in Alloys,*, (ed. Perkins, J.), Plenum Press, 59–87.

[55] Pitteri, M., 1984, *Reconciliation of the local and global symmetries of crystals*, Journal of Elasticity, 14, 175–190.

[56] Pitteri, M., 1985, *On $\nu + 1$-lattices*, Journal of Elasticity, 15, 3–25.

[57] Saburi, T. and Wayman, C.M., 1977, *Crystallographic similarities in shape memory martensites*, Acta Metallurgica, 27, 976–995.

[58] Shimizu, K., 1990, Private communication.

[59] Sohmura, T., Oshima, R. and Fujita, F.E., 1980, *Thermoelastic FCC-FCT transformation in martensitic transformation in Fe-Pd alloys*, Scripta Metallurgica, 14, 855–856.

[60] Subbarao, E.C., 1980, *Zirconia - an overview, Science and Technology of Zirconia* (ed. Heuer, A.H. and Hobbs, L.W.), Advances in Ceramics, 3, 1–24.

[61] Tadaki, T. and Shimizu, K., 1975, *High tetragonality of the thermoelastic martensitic transformation and small volume change during the transformation*, Scripta Metallurgica, 9, 771–776.

[62] Tas, H., Delaey, L. and Deruyttere, A., 1973, *The self-accommodating character of the β_1' copper aluminum martensite*, Metallurgical Transactions, 4, 2833–2840.

[63] Tan, S. and Xu, H., 1990, *Observations on a CuAlNi single crystal*, Continuum Mechanics and Thermodynamics 2, 241–244.

[64] Umemoto, M. and Wayman, C.M., 1978, *Crystallography and morphology studies of Fe-Pt martensites: Lenticular to thin plate transition and thin plate morphologies*, Acta Metallurgica, 26, 1529–1549.

[65] Van Tendeloo, G., 1991, Private communication.

[66] Watanabe, M. and Wayman, C.M., 1971, *Crystallography of the martensite transformation in Fe-Al-C alloys*, Metallurgical Transactions, 2, 2229–2236.

[67] Wechsler, M.S., Lieberman, D.S. and Read, T.A., 1953, *On the theory of the formation of martensite*, Journal of metals, Transactions AIME, 197, 1503–1515.

[68] Xu, H., 1991, Private communication.

[69] Zanzotto, G., 1990, *On the material symmetry group of elastic crystals and the Born rule*, To appear.

APPROXIMATION AND OSCILLATIONS

M. CHIPOT*

1. Introduction. Let Ω be a polygonal domain of $\mathbf{R}^n$ with boundary Γ, $n \geq 1$. We choose Ω polygonal for the simplicity of the numerical analysis. Let $\varphi : \mathbf{R}^n \to \mathbf{R}$ be a Borel function such that there exist distinct W_i's such that

$$\varphi(W_i) = 0 \; \forall\, i = 1, \dots, k \;,\; \varphi(W) > 0 \quad \forall\, W \neq W_i \;,\; i = 1, \dots, k. \tag{1.1}$$

Let A be in the convex hull of the W_i's satisfying

$$A \neq W_i \quad \forall\, i = 1, \dots, k, \tag{1.2}$$

$$A = \sum_{i=1}^{k} \alpha_i W_i \;,\; \alpha_i \in [0,1) \;,\; \sum_{i=1}^{k} \alpha_i = 1. \tag{1.3}$$

Denote by $W_A^{1,\infty}(\Omega)$ the space

$$W_A^{1,\infty}(\Omega) = \{ v \in W^{1,\infty}(\Omega) \mid v(x) = A.x \text{ on } \Gamma \}, \tag{1.4}$$

where $A.x$ is the scalar product of A and x. For $v \in W_A^{1,\infty}(\Omega)$ set

$$E(v) = \int_\Omega \varphi(\nabla v(x))\, dx. \tag{1.5}$$

We will assume all along that

$$\varphi \text{ is bounded on bounded subsets of } \mathbf{R}^n \tag{1.6}$$

in such a way that (1.5) makes sense for every $v \in W^{1,\infty}(\Omega)$. Then we would like to consider the problem

$$\inf_{W_A^{1,\infty}(\Omega)} E(v). \tag{1.7}$$

Different questions arise. Can we compute the infimum (1.7)? Is it achieved? If not can we describe and analyse numerically the minimizing sequences?

A variant of (1.7) could be to consider

$$\hat{E}(v) = \int_\Omega \varphi(\nabla v(x)) + (v - A.x)^2\, dx. \tag{1.8}$$

---

*Université de Metz, Département de Mathématiques, Ile de Saulcy, 57045 METZ-CEDEX 01, (FRANCE)

We refer the reader to [C.C.] for details on this kind of problems introduced in [B.M.].

In the case $n = 2$, a suitable choice for φ could be

$$\varphi(\xi_1, \xi_2) = (\xi_1)^2 + (\xi_2^2 - 1)^2 \tag{1.9}$$

i.e

$$E(v) = \int_\Omega u_x^2 + (u_y^2 - 1)^2 dxdy \tag{1.10}$$

(see for instance [K.M.]). In this case we have

$$W_1 = (0, 1)\ ,\ W_2 = (0, -1) \tag{1.11}$$

and $A.x = 0$ is a suitable boundary data.

In fact, in general one can always assume that $A = 0$ - and we will do so from now on.

Indeed, if $v \in W_A^{1,\infty}(\Omega)$, then $\tilde{v} = v - A \cdot x \in W_0^{1,\infty}(\Omega)$ and

$$\begin{aligned}\int_\Omega \varphi(\nabla v(x))\ dx &= \int_\Omega \varphi(\nabla \tilde{v}(x) + A)\ dx \\ &= \int_\Omega \tilde{\varphi}(\nabla \tilde{v}(x))\ dx\end{aligned} \tag{1.12}$$

where we have set

$$\tilde{\varphi}(W) = \varphi(W + A).$$

Thus, for instance to minimize E on $W_A^{1,\infty}(\Omega)$ is equivalent to minimize the last integral of (1.12) on $W_0^{1,\infty}(\Omega)$. The wells associated with $\tilde{\varphi}$ are $W_i - A$ and they satisfy (1.3) with $A = 0$.

The above problems are scalar version of problems arising in nonlinear elasticity. u is the deformation of the body, Ω the reference configuration. (See [Ci.], [B.J.$_{\cdot 1}$], [B.J.$_{\cdot 2}$], [C.$_{\cdot 1}$], [Er.], [F.$_{\cdot 1}$], [F.$_{\cdot 2}$], [J.$_{\cdot 1}$], [J.$_{\cdot 2}$], [J.K.], [K.], [S]). For simplicity we are restricting ourselves to the scalar case. More general analysis can be found in [C.C.K.]. Previous studies in the one dimensional case have been done in [C.K.L.] and [C.L.$_{\cdot i}$].

2. Energy estimate. First it is clear that it results from relaxation theory that the infimum (1.7) is 0. However, we would like to present here a direct proof of this fact which gives also an additional estimate of (1.7) on some finite element space. For that, let τ be a regular triangulation of Ω. For any simplex $K \in \tau$ we denote by $diam(K)$ the diameter of K. We define

$$h = \text{Max}_{K \in \tau} diam(K)$$

the mesh size. If $P_1(K)$ denotes the space of polynomials of degree less or equal to 1 on K, $v|_K$ the restriction of a function v to K set

$$V_h^0 = \{v : \Omega \to \mathbf{R},\ v \text{ continuous} \mid v|_K \in P_1(K)\ \forall K \in \tau,\ v = 0 \text{ on } \Gamma\}.$$

Then we have :

THEOREM 1. *Under the preceeding assumptions, in particular if (1.6) holds then there exists a constant C such that*

$$\inf_{V_h^0} E(v) \leq Ch^{\frac{1}{2}}. \tag{2.1}$$

Proof. For simplicity we give the proof in the case where E is defined by (1.10) (see [C.$_2$] for the general case).

First note that there is no loss of generality in assuming $h < 1$. Then for any $\alpha \in (0,1)$

$$h^\alpha \gg h$$

when $h \to 0$. Let $\hat{u}_h$ be the function periodic of period $2h^\alpha$ in the y-direction and defined by

$$\hat{u}_h(x,y) = \begin{cases} y \text{ if } y \in [0, h^\alpha] \\ \\ -y + 2h^\alpha \text{ if } y \in [h^\alpha, 2h^\alpha]. \end{cases} \tag{2.2}$$

Clearly $E(\hat{u}_h) = 0$, but $\hat{u}_h$ is not in V_h^0 since it does not necessarily matches the boundary condition, and $\hat{u}_h|_K \notin P_1(K)$ for every K. Then, define u_h as the function of V_h^0 which agrees with

$$\hat{u}_h \wedge dist(x, \Gamma)$$

on the nodes of the triangulation τ ($\wedge$ denotes the minimum of two numbers or two functions). One has

$$\nabla u_h = W_1 \text{ or } W_2 \tag{2.3}$$

except on a neighbourhood of Γ of volume at most $2|\Gamma|h^\alpha$, where $|\Gamma|$ is the measure of Γ, and in a neighbourhood of thickness $2h$ around the lines $y = kh^\alpha$, $k \in \mathbf{Z}$. The number of such lines crossing Ω is of the order of magnitude $diam(\Omega)/h^\alpha$.

Then it results from (1.6) and (2.3) that for some constant

$$E(u_h) \leq C(h^\alpha + h^{1-\alpha}) \leq h^{\alpha \wedge 1-\alpha}. \tag{2.4}$$

Now, $\alpha \wedge 1 - \alpha$ is maximum when $\alpha = 1/2$, which gives (2.1).

Remark 1. We do not know if this estimate is sharp. We refer to [C.L.$_4$], [C.$_2$] for results in this direction.

Remark 2. Our assumptions are extremely weak in terms of φ.

The following theorem precises when the infimum of (2.1) is achieved. One has

THEOREM 2. *Assume that φ is continuous and $\lim_{|\xi| \to +\infty} \varphi(\xi) = +\infty$, then there exists $u_h \in V_h^0$ such that*

$$E(u_h) = \inf_{V_h^0} E(v) \tag{2.5}$$

Proof. For $v \in V_h^0$, $E(v)$ is a continuous function of the values of v at the nodes of the triangulation. This function goes to $+\infty$ when one of the values goes to $+\infty$. The result follows then from a standard compactness argument.

3. Uniqueness of Young measure - Nonexistence of minimizers. Let us assume in this section that the number of wells is less then n. i.e.

$$2 \leq k \leq n. \tag{3.1}$$

Moreover suppose that the vectors

$$W_i - W_1 \ , \ i = 1, \dots , k \text{ are linearly independent.} \tag{3.2}$$

This guarantees that the $\alpha_i \in [0,1)$ such that

$$0 = \sum_{i=1}^{k} \alpha_i W_i \ , \ \sum_{i=1}^{k} \alpha_i = 1 \tag{3.3}$$

are unique.

As we will see - and this is of course supported by computer experiments - minimizing sequences have a common pattern. The pattern of a minimizing sequence is encoded in the Young measure that it generates. Thus, in the case at hand, one expects this Young measure to be unique.

Recall that if u_h is a minimizing "sequence" of (1.7) and if for some constant C

$$|u_h|_\infty \ , \ ||\nabla u_h||_\infty \leq C \tag{3.4}$$

then u_h defines a parametrized measure ν_x on $\mathbf{R}^n$ -so called Young measure- in the sense that there exists a subsequence of u_h, that we still will denote by u_h, and a probability measure ν_x such that for any bounded Carathéodory function $F(x,\xi)$ one has

$$\lim_{h \to 0} \int_\Omega F(x, \nabla u_h(x)) \, dx = \int_\Omega \int_{\mathbf{R}^n} F(x,\lambda) \, d\nu_x(\lambda) dx \tag{3.5}$$

(see [B.], [C.K.], [D.], [Ev.], [K.], [K.P.$_1$], [K.P.$_2$], [P.], [T.], for these questions).

Then we can prove :

THEOREM 3. *Assume φ continuous. There exists a unique Young measure associated to the problems* (1.7) *which is given by*

$$\nu_x = \sum_{i=1}^{k} \alpha_i \delta_{W_i} \tag{3.6}$$

where δ_{W_i} denotes the Dirac measure at W_i, α_i the parameters of (3.3).

Proof. First it results from the construction given in section 2 that there exist minimizing sequences satisfying (3.4) - thus there exist Young measures for (1.7).

Let u_h be a minimizing sequence of (1.7) such that (3.4) holds.

Up to a subsequence one can assume that one has for some $u \in W^{1,\infty}(\Omega)$

$$u_h \to u \text{ uniformly in } \Omega \tag{3.7}$$

$$\nabla u_h \rightharpoonup \nabla u \quad \text{in } L^\infty(\Omega) \quad \text{weak} - *. \tag{3.8}$$

((3.8) means that each component of the gradient is converging in $L^\infty(\Omega)$ weak$-*$).
Let ν_x be a Young measure defined by u_h.
From (3.5) applied with $F(x,\xi) = \varphi(\xi)$ we have

$$0 = \lim_{h \to 0} \int_\Omega \varphi(\nabla u_h(x))\, dx = \int_\Omega \int_{\mathbf{R}^n} \varphi(\lambda)\, d\nu_x(\lambda) dx.$$

It results that

$$\int_{\mathbf{R}^n} \varphi(\lambda)\, d\nu_x(\lambda) = 0 \quad \text{a.e. in } \Omega.$$

Hence, by (1.1) the support of ν_x has to be contained in the W_i's for a.e. x, i.e.

$$\nu_x = \sum_{i=1}^{k} \beta_i(x) \delta_{W_i}. \tag{3.9}$$

By (3.1), (3.3) we see that the W_i span a proper subspace of $\mathbf{R}^n$. Let us denote by ν a unit vector in the orthogonal complement of the span of the W_i's. If we apply (3.5) with

$$F(x,\xi) = |\xi.\nu|$$

we obtain

$$\int_\Omega |\nabla u_h(x).\nu| dx \to \int_\Omega \int_{\mathbf{R}^n} |\lambda.\nu| d\nu_x(\lambda) dx = 0$$

by (3.9). It results that

$$\frac{\partial}{\partial \nu} u_h \to 0 \text{ in } L^1(\Omega).$$

Then by the Poincaré inequality

$$u_h(x) \to 0 \text{ in } L^1(\Omega).$$

Since this is true for any subsequence of u_h it follows that we have when $h \to 0$

$$\nabla u_h \rightharpoonup 0 \text{ in } L^\infty(\Omega) \quad \text{weak} - *.$$

Let B be a ball included in Ω.
From (3.5) applied with $F(x,\xi) = X_B(x)\xi$ (this function is not bounded in ξ, but the gradient of u_h is) we obtain

$$0 = \lim_{h \to 0} \int_B \nabla u_h(x)\, dx = \int_B \int_{\mathbf{R}^n} \lambda\, d\nu_x(\lambda) dx. \tag{3.10}$$

By (3.9) this reads

$$0 = \sum_{i=1}^{k} \int_B \beta_i(x)\, dx \cdot W_i.$$

Hence, by (3.3) and the uniqueness of barycentric coordinates

$$\frac{1}{|B|} \int_B \beta_i(x)\, dx = \alpha_i \quad \forall\, i = 1, \ldots, k.$$

(Recall that since ν_x is a probability measure $\sum_{i=1}^k \beta_i(x) = 1$). Since this is true for any B it results from the Lebesgue differentiation theorem that

$$\beta_i(x) = \alpha_i \quad \text{a.e. in } \Omega, \quad \forall\, i = 1, \ldots, k$$

and thus (3.9) reads

$$\nu_x = \sum_{i=1}^{k} \alpha_i \delta_{W_i} \quad \text{a.e. in } \Omega.$$

This shows that ν_x is necessarily the homogeneous Young measure (3.6) (see [K.P.$_1$]) and concludes the proof.

A consequence of this is:

THEOREM 4. *Under the assumptions of Theorem 3, the problem* (1.7) *does not admit a minimizer.*

Proof. Let $u \in W_0^{1,\infty}(\Omega)$ be a minimizer to (1.7). Then, $\nu_x = \delta_{\nabla u(x)}$ the Dirac mass at the point $\nabla u(x)$ is a Young measure for (1.7). By the uniqueness theorem that we just proved we must have

$$\delta_{\nabla u(x)} = \sum_{i=1}^{k} \alpha_i \delta_{W_i} \text{ a.e. } x \in \Omega. \tag{3.11}$$

But this imposes

$$\nabla u(x) = W_i \ , \ \alpha_i = 1 \text{ a.e. } x \in \Omega$$

which is impossible by (1.2).

Remark 3. It should be noticed that one cannot remove the assumption (3.1). For instance if on some triangle Ω in $\mathbf{R}^2$ we consider the "hat" function u that vanishes on the edge of this triangle and is equal to 1 at some point inside - and of course interpolates linearly between these values - then the gradient of this function takes three values say

$$W_1,\ W_2,\ W_3.$$

Then if we choose φ such that

$$\varphi(W_i) = 0\ \forall\, i = 1, 2, 3$$

this hat function is a minimizer of E over $W_0^{1,\infty}(\Omega)$. Moreover, By the divergence theorem one has

$$0 = \frac{1}{|\Omega|} \int_\Omega \nabla u(x) dx = \sum_{i=1}^{3} \alpha_i W_i$$

which is (1.3).

4. Numerical analysis of oscillations. The uniqueness of the Young measure associated to (1.7) is an indication of some coherence in the oscillations of the minimizing sequences. This is what we would like to investigate now.

For this let

$$0 < R < \frac{1}{2}|W_i - W_j| \quad \forall\, i \neq j.$$

Let B be a ball included in Ω. For $v \in H^1(\Omega)$, set

$$B_i^R(v) = B_i^R = \{x \in B \mid |\nabla v(x) - W_i| < R\}.$$

If $|\ |$ denotes the Lebesgue measure, then $|B_i^R(v)|/|B|$ is the probability for v to have its gradient in the ball of center W_i and radius R. Let us denote by $B_{ex}^R(v)$

$$B_{ex}^R(v) = B_{ex}^R = B \setminus \cup_i B_i^R.$$

Assume that there exists a constant Λ such that for $p > 1$

$$\varphi(\xi) \geq \Lambda . \min_i |\xi - W_i|^p. \tag{4.1}$$

Note that in the example (1.9) one has

$$\varphi(\xi_1, \xi_2) \geq \min(\xi_1^2 + (\xi_2 + 1)^2, \xi_1^2 + (\xi_2 - 1)^2) = \min(|\xi - W_1|^2, |\xi - W_2|^2) \tag{4.2}$$

Then we have

THEOREM 5. *Assume that φ satisfies (4.1). There exists a constant C such that*

$$\left|\frac{|B_i^R(v)|}{|B|} - \alpha_i\right| \leq C.\{E(v)^{\frac{1}{pp'}} + E(v)^{\frac{1}{p}}\} \tag{4.3}$$

$$|B_{ex}^R(v)| \leq C.E(v) \tag{4.4}$$

$(\frac{1}{p} + \frac{1}{p'} = 1)$.

Proof. One has

$$\begin{aligned} |B_{ex}(v)| &\leq \frac{1}{R^p} \int_\Omega \min_i |\nabla v(x) - W_i|^p dx \\ &\leq \frac{1}{R^p} E(v) \end{aligned}$$

which proves (4.4). We refer the reader to [C.2] for the proof of (4.3).

As a consequence we see that if v_h is some function of V_h^0 such that

$$E(v_h) \leq 2. \inf_{V_h^0} E(v)$$

then by Theorem 1 one has, when h is small:

$$\left|\frac{|B_i^R(v_h)|}{|B|} - \alpha_i\right| \leq C.h^{\frac{1}{2pp'}}.$$

In the particular case of example (1.10) we have

$$\left|\frac{|B_i^R(v_h)|}{|B|} - \alpha_i\right| \leq C.h^{\frac{1}{8}}.$$

This gives an estimation in terms of h of the probability for v_h to have its gradient in a neighbourhood of W_i. This probability approaches α_i when $h \to 0$.

We would like to end this section by some numerical results. Let us assume that Ω is the square $(-1,1)^2$, and consider the case of the two wells

$$W_1 = (-1,1) \quad , \quad W_2 = (1,-1).$$

Moreover let us choose

$$\varphi(\xi) = |\xi - W_1|^2 \ |\xi - W_2|^2.$$

It is not difficult to check that these data fit our previous analysis. Since

$$0 = \frac{1}{2}W_1 + \frac{1}{2}W_2$$

the minimizing sequences of (1.7) are picking up their gradients alternatively around W_1 and W_2 with a probability which tends to $\frac{1}{2}$ (see Theorem 5). This can be observed on the figure below :

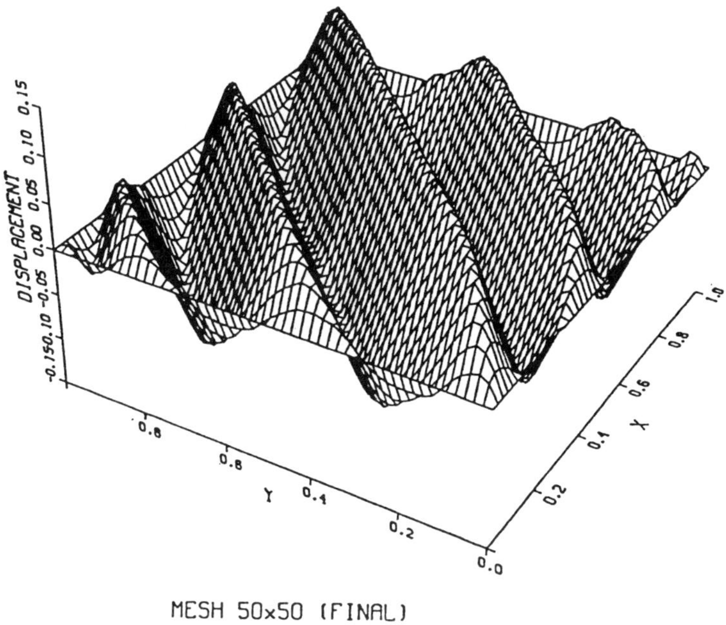

Computational Result for $h = 1/50$.

In fact, in order to lower the energy, the computer uses almost exactly the technique developed in Theorem 1. The oscillations of the minimizing sequences are stopped at the level of the grid.

Our numerical codes use standard iterations algorithms. We start with a random distribution of initial values of the function close to 0 and we decrease the energy using a conjugate gradient method. An example of the final result is displayed on the above figure.

The most interesting examples of applications of our results occur in the study of deformations of ordered materials as crystals. In this case the potential wells are matrices. Moreover, due to the frame indifference, these wells are in infinite number. Also deformations have to satisfy the well known compatibility condition. This makes the problem more involved. For some results in this direction and complements we refer the reader to [C.C.K.] and [Co.].

Acknowledgements. This work was done during one visit at the University of Minnesota. I would like to thank the I.M.A. and the School of Mathematics who are providing such a good working atmosphere. This research has been supported there by the NSF and ASOFR through DMS 87-18881.

I thank also C. Collins and D. Kinderlehrer for their collaboration.

REFERENCES

[B.] J. M. Ball, *A version of the fundamental theorem for Young measures*, in *Partial Differential Equations and Continuum Models of Phase Transitions*, Lecture Notes in Physics, # 344 M. Rascle, D.Serre and M. Slemrod, eds., Springer-Verlag, 1989, pp. 207–215.

[B.J.$_1$] J. M. Ball and R. D. James, *Fine phase mixtures as minimizers of energy*, Arch. Rational Mech. Anal., 100 (1987), pp. 13–52.

[B.J.$_2$] J. M. Ball and R. D. James, *Proposed experimental tests of a theory of fine microstructure and the two wells problem*, preprint (1990).

[B.M.] J. M. Ball and F. Murat, *$W^{1,p}$ quasiconvexity and variational problems for multiple integrals*, J. Funct. Anal., 58 (1984), pp. 225–253.

[C.$_1$] M. Chipot, *Hyperelasticity for crystals*, European Journ. of Appl. Math., 1 (1990), pp. 113–129.

[C.$_2$] M. Chipot, *Numerical analysis of oscillations in nonconvex problems*, (to appear).

[C.C.] M. Chipot and C. Collins, *Numerical approximation in variational problems with potential wells*, preprint (1990) (to appear).

[C.C.K.] M. Chipot, C. Collins and D. Kinderlehrer, (In preparation).

[C.K.] M. Chipot and D. Kinderlehrer, *Equilibrium configurations of crystals*, Arch. Rational Mech. Anal., 103 (1988), pp. 237–277.

[Ci.] P.G. Ciarlet, *Elasticité tridimensionnelle*, Masson, 1985.

[Co.] C. Collins, *Thesis, University of Minnesota*, 1990.

[C.K.L.] C. Collins, D. Kinderlehrer and M. Luskin, *Numerical approximation of the solution of a variational problem with a double well potential*, IMA preprint, # 605 (1990); (To appear in SIAM J. Numer. Anal.).

[C.L.$_1$] C. Collins and M. Luskin, *The computation of the austenitic-martensitic phase transition*, in *Partial Differential Equations and Continuum Models of Phase Transitions*, Lecture Notes in Physics # 344, M. Rascle, D.Serre and M. Slemrod, eds., Springer-Verlag, Berlin, New York, 1989, pp. 34–50.

[C.L.$_2$] C. Collins and M. Luskin, *Computational results for phase transitions in shape memory materials.*, in *Smart Materials, Structures, and Mathematical Issues*, C. Rogers, ed. Technomic Publishing Co., Lancaster, Pennsylvania, (1989), pp. 198–215.

[C.L.$_3$] C. Collins and M. Luskin : Numerical modeling of the microstructure of crystals with symmetry-related variants, in *Proceedings of the ARO US-Japan Workshop on Smart/Intelligent Materials*, Technomic Publishing Co., Lancaster, Pennsylvania (to appear).

[C.L.$_4$] C. Collins and M. Luskin, *Optimal order error estimates for the finite element approximation of the solution of a nonconvex variational problem*, Preprint (1990).

[D.] B. Dacorogna, *Weak continuity and weak lower semicontinuity of nonlinear functionals*, Springer Lecture Notes # 922, 1982.

[Er.] J. L. Ericksen, *Some constrained elastic crystals.*, in *Material Instabilities in Continuum Mechanics and Related Problems*, J. M. Ball, ed., Oxford University Press, Oxford, 1987, pp. 119–137.

[Ev.] L. C. Evans, *Weak Convergence Methods for Nonlinear Partial Differential Equations*, A.M.S. Regional Conference Series in Mathematics, # 74 (1989).

[F.$_1$] I. Fonseca, *Variational methods for elastic crystals.*, Arch. Rat. Mech. Anal., 97 (1985), pp. 189–220.

[F.$_2$] I. Fonseca, *The lower quasiconvex envelope of the stored energy function for an elastic crystal*, J. Math. Pures et Appl., 67 (1988), pp. 175–195.

[G.T.] D. Gilbarg and N.S. Trudinger, *Elliptic Partial Differential Equations of Second Order*, Springer Verlag, 1985.

[J.$_1$] R. James, *Basic principles for the improvement of shape-memory and related materials*, in *Smart Materials, Structures, and Mathematical Issues*, C. Rogers, ed., Technomic Publishing Co. Lancaster, PA, 1989.

[J.$_2$] R. D. James, *Microstructure and weak convergence.*, in *Material Instabilities in Continuum Mechanics and Related Problems*, J. M. Ball, ed., Oxford University Press, Oxford, 1987, pp. 175–196.

[J. K.] R. D. James and D. Kinderlehrer, *Theory of diffusionless phase transitions, Partial Differential Equations and Continuum Models of Phase Transitions*, Lecture Notes in Physics # 344, M. Rascle, D.Serre and M. Slemrod, eds., Springer-Verlag, Berlin, New York, 1989, pp. 51–84.

[K.] D. KINDERLEHRER, *Remarks about equilibrium configurations of crystals*, in *Material Instabilities in Continuum Mechanics and Related Problems*, J. M. Ball ed., Oxford University Press, Oxford, 1987, pp. 217–242.

[K. P.$_1$] D. KINDERLEHRER AND P. PEDREGAL, *Analysis of Young measures*, Preprint.

[K. P.$_2$] D. KINDERLEHRER AND P. PEDREGAL, *Weak convergence of integrands and the Young measure representation*, Preprint.

[Ko.] R. KOHN, *The relationship between linear and nonlinear variational models of coherent phase transitions.*, in *Proceedings of the Seventh Army Conference on Applied Mathematics and Computing*, West Point, June 1989.

[K.M.] R. KOHN AND S. MÜLLER, (To appear).

[P.] P. PEDREGAL, Thesis, University of Minnesota (1989).

[S.] S. A. SILLING, *Phase changes induced by deformation in isothermal elastic crystals*, preprint (1988).

[T.] L. TARTAR, *Compensated compactness and application to partial differential equations*, in *Nonlinear analysis and mechanics: Heriot–Watt Symp IV*, R.J. Knops ed., Pitman, 1979, pp. 136–212.

COMPUTATION OF TWINNING

CHARLES COLLINS*

Introduction. In studying twinning in crystals one is lead to studying combinations of gradients which come from the potential wells of the material energy. Our approach to determining these possible microstructures is to form a particular energy and to numerically construct approximate minimizers. In this paper, we give a brief introduction to the two-well problem and our computational approach to solving it. Then we present and discuss our results for two and more variant microstructures.

Two-well problem. First we give a brief introduction to the two-well problem; for more detailed information see [BJ2], and for further information on twinning in general see [BJ1],[Ch],[CK],[JK] and [K]. For the two-well problem we have an energy which has two potential wells and we wish to investigate all possible microstructures constructable from variants of these two wells. The deformation gradients U_0 and U_1 minimize the material energy and, since we assume rotational invariance, the wells consist of all rotations of these matrices. So we have the U_0-well and the U_1-well with

$$U_i\text{-well} = \{RU_i \colon R \text{ is a rotation}\}.$$

A variant is a particular element of one of the wells. We assume also that the gradients U_0 and U_1 are symmetry related, that is, there exists a matrix H, so that $U_0 = U_1 H$. This assumption puts this description in line with the theory of martensitic twinning. A microstructure is an arrangement of variants and other gradients in such a way that the resulting deformation is continuous and when we scale the microstructure down only the variants participate in the limit. In this way we produce energy minimizing sequences. This restriction to continuous deformations limits which variants may participate in each configuration. The microstructures of the two-well problem are two-dimensional and so we can use a two-dimensional model, making the computations easier.

The bulk energy associated with a deformation, $y(x)$, of the reference domain, $\Omega = [0,1]^2$, is given by

$$J(y) = \int_\Omega W(\nabla y(x))\, dx$$

where $W(F)$ is the energy density. For our two-dimensional energy model we take $W(F) = \phi(F^T F)$ with $C = F^T F$ and

$$\phi(C) = \kappa_1(\operatorname{tr} C - 2)^2 + \kappa_2 C_{12}^2 + \kappa_3\left(\left(\frac{C_{11} - C_{22}}{2}\right)^2 - \epsilon^2\right)^2 \tag{1}$$

*Department of Mathematics, University of Michigan, Ann Arbor, MI 48109-1003

with constants $\kappa_i > 0$ and $\epsilon > 0$ given. The form of the energy is based on a suggestion by Ericksen and James and is based on Ericksen's work on constrained elastic crystals ([E1-2]). This energy has two minimizing deformation gradients,

$$U_0 = \begin{pmatrix} \eta_1 & 0 \\ 0 & \eta_2 \end{pmatrix}, \text{ and } U_1 = \begin{pmatrix} \eta_2 & 0 \\ 0 & \eta_1 \end{pmatrix}, \tag{2}$$

with $\eta_1 = \sqrt{1-\epsilon}$ and $\eta_2 = \sqrt{1+\epsilon}$. These two deformation gradients are symmetry related by $U_1 = U_0 H$ where $H = R(\frac{\pi}{2})$. The form of the wells agrees with previous work on the austenite-martensite interface ([BJ1], [CL1-2] [JK]).

To solve the problem we form a set of admissible deformations and then minimize the bulk energy over this set. For our possible deformations we consider continuous functions which satisfy the boundary condition $y(x) = Fx$, where F is some appropriately chosen matrix.

For our numerical computations we use continuous, bilinear finite elements on a uniform square mesh of size h. We use an optimization algorithm based on the Polak-Ribeŕe version of the conjugate gradient algorithm [C]. For initial conditions we take a perturbation of the homogeneous deformation $y(x) = Fx$. See [A] and [S] for other computational approaches to similar problems. See [CC], [CKL] and [CL3] for some justification for this numerical approach.

Two-variant twinning. The first microstructure we wish to investigate is two-variant twinning. We know that there are no mixtures available within one well, so we expect a banded structure using one variant from each well. To construct this we determine a rotation R and vectors a and n such that $RU_1 = U_0 + a \otimes n$ and then establish boundary conditions compatible with a banded microstructure. We have

$$R = \begin{pmatrix} \eta_1\eta_2 & \pm\epsilon \\ \mp\epsilon & \eta_1\eta_2 \end{pmatrix},$$

$n = \frac{1}{\sqrt{2}}(1, \pm 1)$ and $a = \sqrt{2}\epsilon(\eta_1, \mp\eta_2)$. We take $n = \frac{1}{\sqrt{2}}(1,1)$ and form the average deformation gradient

$$\begin{aligned} F_\lambda &= (1-\lambda)U_0 + \lambda R U_1 \\ &= \begin{pmatrix} \eta_1(1+\lambda\epsilon) & \eta_1\lambda\epsilon \\ -\eta_2\lambda\epsilon & \eta_2(1-\lambda\epsilon) \end{pmatrix}. \end{aligned} \tag{3}$$

We have $W(F_\lambda) = 4(\kappa_1 + 4\kappa_2 + \kappa_3)\epsilon^4\lambda^2(1-\lambda)^2$ so F_λ is not a minimizer of the energy for $0 < \lambda < 1$. We induce this martensitic-type twinning by using the displacive boundary condition, $y(x) = F_\lambda x$. By [BJ2] the Young measure derived from any minimizing sequence satisfying this boundary condition is unique and has the form

$$\nu_x \equiv \nu = (1-\lambda)\delta_{F_0} + \lambda\delta_{F_1}.$$

Thus we would expect our approximate minimizers to attain the fine band structure typical of martensitic twinning with the ratio of the two variants depending on the parameter λ.

Figure 1. Two-variant twinning. ($\lambda = 1/2$, $h = 1/64$)

To visualize our results we use the function

$$\psi(F) = \frac{\|F^T F - U_0^2\|}{\|F^T F - U_0^2\| + \|F^T F - U_1^2\|},$$

where F is a deformation gradient and $\|A\|^2 = \sum A_{ij}^2$. We shade the elements of the reference domain by the value of $\psi(F)$ for the gradient in that element; white corresponds to 0, black to 1 and shades of gray to values in between 0 and 1. We call this the colored picture.

We have performed numerous numerical experiments using different value of the parameters κ_i and the numerical methods used (see [C]). In Figure 1 we see the result for the case $\lambda = 1/2$, $h = 1/64$, $\kappa_1 = 10$, $\kappa_2 = 3$, $\kappa_3 = 1$ and $\epsilon = 0.1$. We obtain twinning on the scale of the mesh. For color video of this approximation see [CLR]. With other values of λ we see different ratios for the variants used and different widths for the bands.

Mesh considerations/rotated problem. In our problem we have three forces participating in the computation: preserving continuity, maintaining the boundary condition and attaining minimum energy. In typical finite-element problems the mesh's role is to pass information from these various parts in an efficient way. In our case, however, the mesh participates and becomes part of the problem in terms of preserving continuity. Its effect is something which could be associated with a non-isotropic surface energy.

With a uniform square mesh we have a reference frame in which we can express all possible configurations locally. In addition we assign the value of the gradient in each element to the value at the midpoint which gives more freedom in the choice of adjoining gradients. In this way we have reduced the effect of the mesh on the computations. However, we have sacrificed the ability to resolve the features clearly and have biased our results towards homogeneous deformations.

From [CL1-2], we know that for some problems an improper mesh orientation can make it very difficult if not impossible to twin. One solution to this problem is to rotate the entire problem so we can adjust the twin plane normal with respect to the mesh. We also use the various rotations to test different orientations of the mesh.

For our change we denote the new rotated quantities with hats. Given a rotation R, we transform the domain from Ω to $\widehat{\Omega} = R\Omega$ by $\widehat{x} = Rx$. We also rotate the deformed space so $\widehat{y} = Ry$. This gives us $\widehat{y}(\widehat{x}) = Ry(R^T\widehat{x})$ and

$$\widehat{\nabla}\widehat{y}(\widehat{x}) = R\nabla y(R^T\widehat{x})R^T.$$

Thus every gradient matrix F in the original problem is replaced by RFR^T and we have

$$\begin{aligned}\widehat{W}(\widehat{F}) &= W(R^T\widehat{F}R)\\ \widehat{F}_\lambda &= RF_\lambda R^T\\ &= RU_0R^T + \lambda Ra \otimes Rn.\end{aligned}$$

Thus the new twin plane normal is Rn, and we can choose the angle of rotation so that $\widehat{n} = Rn$ is in any direction we want. In the actual implementation we apply the inverse rotation to the rotated version of the problem and move everything back to the original domain. Thus, the domain is still $[0,1]^2$ and the finite element mesh still prefers the orientations $(1,\pm1)$, $(1,0)$ and $(0,1)$. This allows us to choose $\widehat{n}$ so that it is not compatible with the mesh grid lines. The analytic conclusions for the rotated problem remain the same as for the original problem with just a change in the matrices involved.

Computationally, we still are able to have two-variant twinning using various angles (see Figure 2). The quality of the twinning changes, because the rotated variants and the preferred banded structure are not compatible with the mesh.

More than two-variant twinning. We now want to look at other possible microstructures constructable from variants on two wells. We have already exhausted the case when we use two variants, so the next type of microstructure we wish to examine is a three-variant mixture. As before, the microstructure is banded, but now two of the variants twin within one of the bands. In this case the rank-one connection is between the two-variant mixture F_λ and the U_0-well. As before, we determine vectors b_λ and m_λ and a rotation Q_λ so that $Q_\lambda U_0 = F_\lambda + b_\lambda \otimes m_\lambda$ and get

$$Q_\lambda = \frac{1}{\eta_1^2\eta_2^2 + \lambda^2\epsilon^2}\begin{pmatrix}\eta_1^2\eta_2^2 - \lambda^2\epsilon^2 & 2\lambda\epsilon\eta_1\eta_2\\ -2\lambda\epsilon\eta_1\eta_2 & \eta_1^2\eta_2^2 - \lambda^2\epsilon^2\end{pmatrix}$$

Figure 2. Two-variant twinning in 30° problem ($\lambda = 1/2$, $h = 1/64$)

and $m_\lambda = \frac{1}{\sqrt{1+\gamma^2}}(-1, \gamma)$, where $\gamma = \frac{\eta_2^4 - \lambda^2\epsilon^2}{\eta_1^2\eta_2^2 + 2\lambda\epsilon + \lambda^2\epsilon^2} \approx 1 + 2\epsilon(1-\lambda)$. And we form the average

$$\begin{aligned} F_{\lambda,\mu} &= (1-\mu)F_\lambda + \mu Q_\lambda U_0 \\ &= \begin{pmatrix} \eta_1(1+\lambda\epsilon(1-\mu-2\lambda\mu\epsilon/M)) & \eta_1\lambda\epsilon(1-\mu+2\mu\eta_2^2/M) \\ -\eta_2\lambda\epsilon(1-\mu+2\mu\eta_1^2/M) & \eta_2(1-\lambda\epsilon(1-\mu+2\lambda\mu\epsilon/M)) \end{pmatrix} \end{aligned} \tag{4}$$

where $M = \eta_1^2\eta_2^2 + \lambda^2\epsilon^2$. For this microstructure we take as our boundary condition $y(x) = F_{\lambda,\mu}x$. We note that $Q_\lambda U_0$ is rank-one related to the average F_λ, but not to either of the components of F_λ, thus with limited resolution, as we have in our computations, some portion of the domain will be used to form the interface between $Q_\lambda U_0$ and a fine mixture of U_0 and RU_1.

Now we have established a pattern for creating further, more complicated microstructures and their corresponding boundary conditions. For example, we can replace a band of the three-variant mixture with a band containing a twinned mixture. We then solve new equations to determine a new set of compatible variants and use the boundary condition $y(x) = Fx$ where F is the proper weighted average of the variants. Repeating this process, we form a set of matrices F which are the boundary conditions corresponding to all the energy minimizing microstructures for the two-well problem. Since any F can be rotated and still give the same type of microstructure, we look at the set

$$\mathcal{M} = \{F^T F \mid F \text{ is the boundary condition for a microstructure}\}. \tag{5}$$

$\mathcal{M}$ is spanned by $\{F_{\lambda,\mu}^T F_{\lambda,\mu}\}$, but this is not a unique description ([BJ2]). Except for the two-variant mixtures, any element of $\mathcal{M}$ can be written as a combination of any number of variants. The domain is a small slice of a shell. Its dimensions are $\epsilon \times \epsilon^2$ and its boundary is formed by the two-variant mixtures (for pictures of $\mathcal{M}$ see [BJ2], [C]). Every matrix in $\mathcal{M}$ is within $\epsilon^2/2$ of the boundary, which causes some computational problems. For instance, there exists a rotation R such that

$$F_{\lambda,\mu} = RF_{\lambda(1-\mu)} + \mathcal{O}(\epsilon^2) \tag{6}$$

so we might observe this improper two-variant microstructures even when we have supplied three-variant type boundary conditions. Thus the form of the energy and the mesh should play an important role in determining which microstructure is obtained computationally.

Returning to the case of boundary conditions of the form $y(x) = F_{\lambda,\mu}x$, we expect twins within twins. From the discussion above we know that this boundary condition is also compatible with a four-variant case where both bands are twinned. For the original, non-rotated problem the resulting minimum energy configuration does not attain bands within bands but takes on the improper two-variant configuration mentioned above (6). Another reason for this improper choice of microstructure may be that as the rank-one connection is between $Q_\lambda U_0$ and F_λ and so for any fixed width approximation of F_λ some adjustments have to be made to interface with $Q_\lambda U_0$. Not paying the penalty of these adjustments between the bands may make up for the penalty of the larger adjustment needed to match the boundary condition.

Since two-variant twinning is still possible within the rotated problem we now consider the case of multivariant twinning in the rotated problem. By rotating the domain we raise the energy needed to form the nearby, but improper, two-variant mixture enough that the expected three (or four or more) variant mixture is now the low energy configuration. By rotating the problem we also increase the energy for forming the interface between the variants and so the resulting minimizing configuration should depend on the angle of rotation used.

Before we present our results, we will first discuss what we expect to see. A minimizing sequence is associated with a Young measure of the form

$$\nu_x = \sum_i \lambda_i(x)\delta_{A_i}$$

where $0 \le \lambda_i(x) \le 1$, $\sum_i \lambda_i = 1$ and the A_i's are variants. We also have

$$\int_\Omega \lambda_i(x)\,dx = \alpha_i|\Omega|$$

where the values α_i are determined by the boundary condition and the choice of the variants A_i used (not necessarily unique). We can then consider the function

$$F(x) = \sum_i \lambda_i(x)A_i$$

which is the value of the gradient at each point x. Under the conditions listed, if we plot the values of $F(x)$ we would expect to see a certain characteristic shape

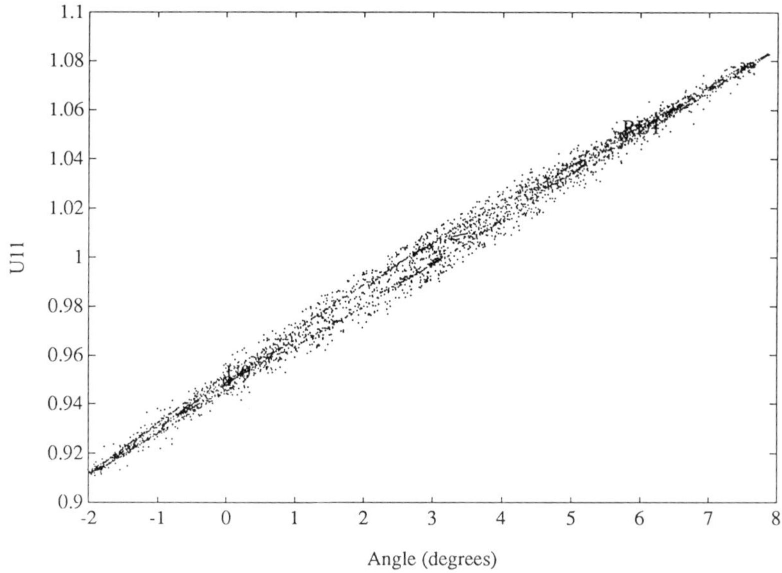

Figure 3. Gradient distribution for two-variant twinning ($\lambda = 1/2$, $h = 1/64$, no rotation).

depending on the number and position of the variants A_i. If there are only two variants then the graph would be a line connecting them, three a triangle, four a parallelogram, etc. There are two other factors affecting the distribution; first, in our approximations we expect $\lambda_i(x)$ to be the characteristic function for some set S_i. Second, we are trying to attain minimum energy so there may be certain paths in this shape which are preferred, in particular rank-one connections between the variants A_i. To judge our approximations we graph the distribution of the gradient values taken. We project the 4-dimensional gradients into 2-dimensions by decomposing the gradient into a rotation R and a symmetric positive definite matrix U, and then plotting U_{11} vs. θ where θ is the angle for the rotation R. This corresponds to a picture of the wells with lines representing the U_0-well (corresponding to $U_{11} = \sqrt{1-\epsilon}$) and the U_1-well (corresponding to $U_{11} = \sqrt{1+\epsilon}$). Any variant of either well is associated with a point on one of these lines, its position determined by the rotation used to obtain it, and rank-one connections can be represented by lines connecting the wells.

When we graph our results in this way, we expect to see certain features. There will be the characteristic shape which will tell us how many variants are involved, and then there will be certain clumping to give us an idea as to what the approximate ratios of the variants are and what low energy connecting paths are used. We can only use this graph if the deformation only contains one microstructure, otherwise we may have two structures interfering with each other. We also expect that as more variants are involved the clumping will be less distinct as more elements will be dedicated to the interface between the bands of variants. In Figure 3 we see the gradient distribution for the original problem with $\lambda = 1/2$, showing the characteristic shape for two variants.

Now we present the results for three angles of rotation. In each case we have $\lambda = \mu = 1/2$ and the parameters from the two-variant case. We show both the gradient distribution and the colored picture, so we can see which variants are involved in the twinning and what the geometric configuration is.

The first case is a 10° rotation as seen in Figure 4. The distribution shows two straight lines, representing two-variant mixtures. Since the line is off the marks it is the improper two-variant twinning seen in the case without any rotation (6). The colored picture indicates that there is little organized structure. The only benefit of the rotation is that a larger region of gradients are involved. This could be considered an intermediate state, where the improper two-variant mixture still is lower in energy but the other options are coming into play. For many other angles of the rotation similar results hold. This is simply a limitation of the present formulation of the problem and the orientation of the mesh.

Next we consider a 30° rotation as seen in Figure 5. Here we see a characteristic triangular shape indicating three-variant twinning. The marks indicate that the correct variants are used for this microstructure. The clump in the middle is due to the gradients forming the boundary layer. In the colored picture we can actually see twins within twins with twin-plane normals of 15° and 102°. The other band is a numerical effect due to the inability of the mesh to completely resolve the non-standard angle of m_λ.

Finally we consider a 45° rotation as seen in Figure 6. The gradients form a characteristic rectangular shape but the clumping leads us to believe that this is not just four-variant twinning. Either we have more than four variants or there are two different microstructures involved in the approximation. It is also possible that there are some mesh effects involved since a 45° rotation strongly prefers certain orientations. We also see here many gradients participating in the interface between the various variants.

Conclusions. Numerical computations are a useful tool for studying the possible microstructure in the two-well problem. Hopefully these and future results can contribute to the understanding and improvement of other approaches to this challenging problem. Although we can obtain results for two-variant and multivariant twinning it is clear that there are inadequacies in the formulation of the problem and the computational details that interfere with a complete understanding of this problem.

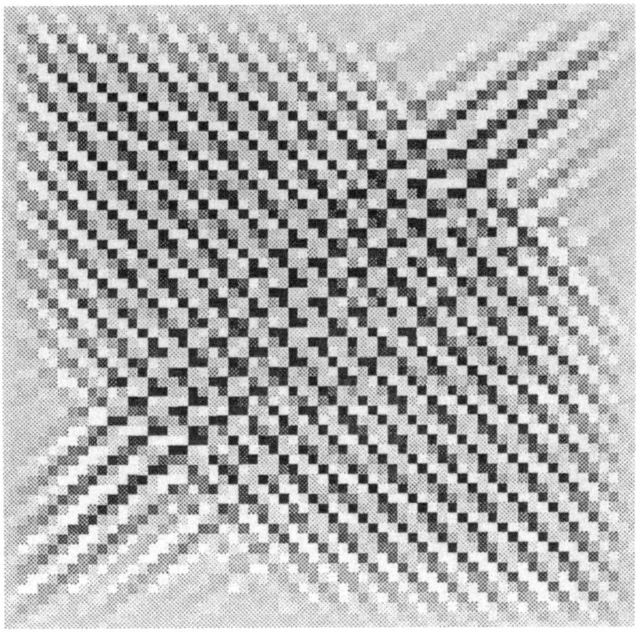

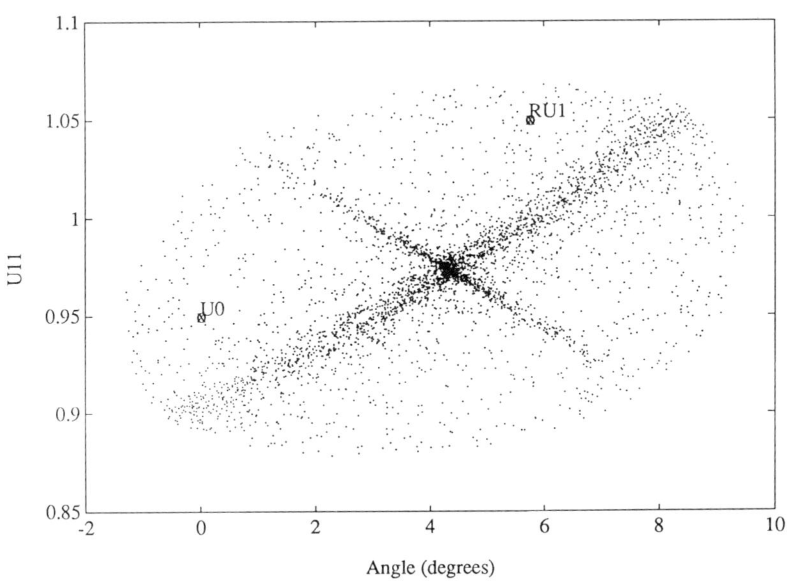

Figure 4. Variant map (top) and gradient distribution (bottom) for 10° problem ($\lambda = 1/2$, $h = 1/64$)

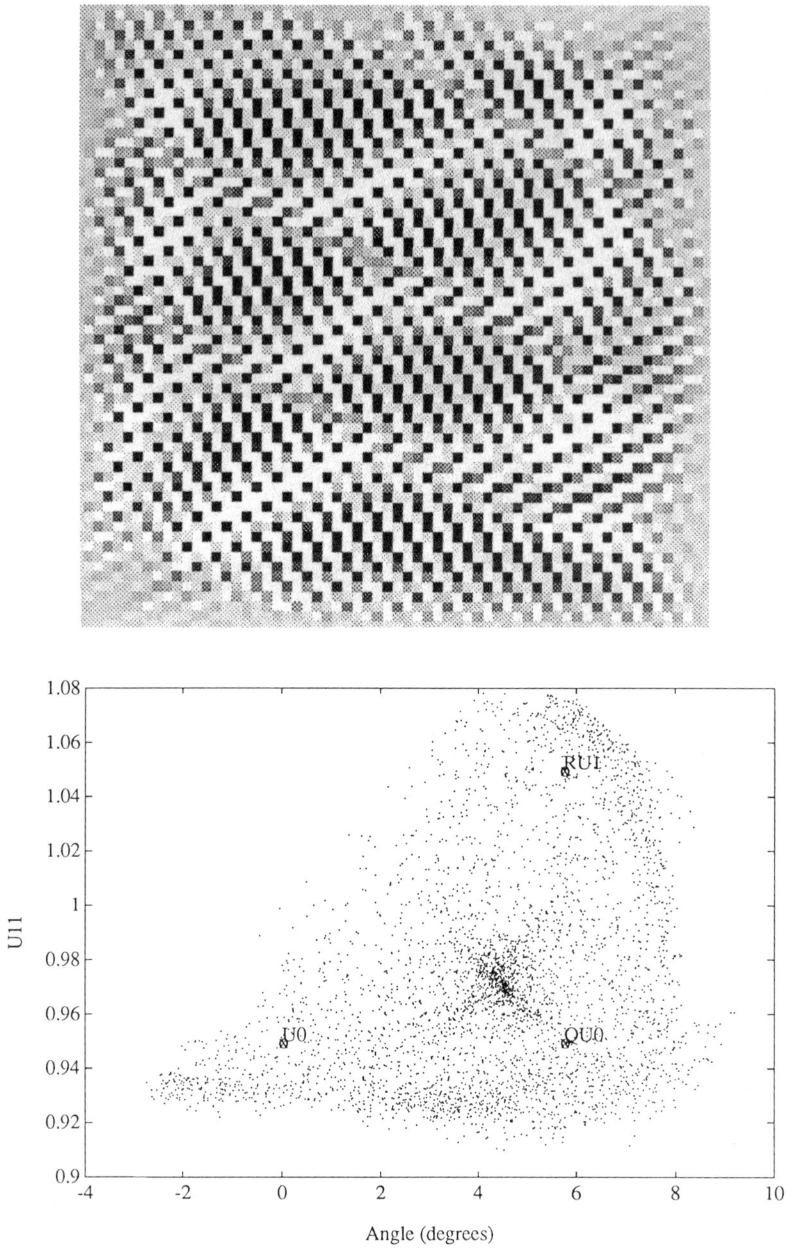

Figure 5. Variant map (top) and gradient distribution (bottom) for 30° problem ($\lambda = 1/2$, $h = 1/64$)

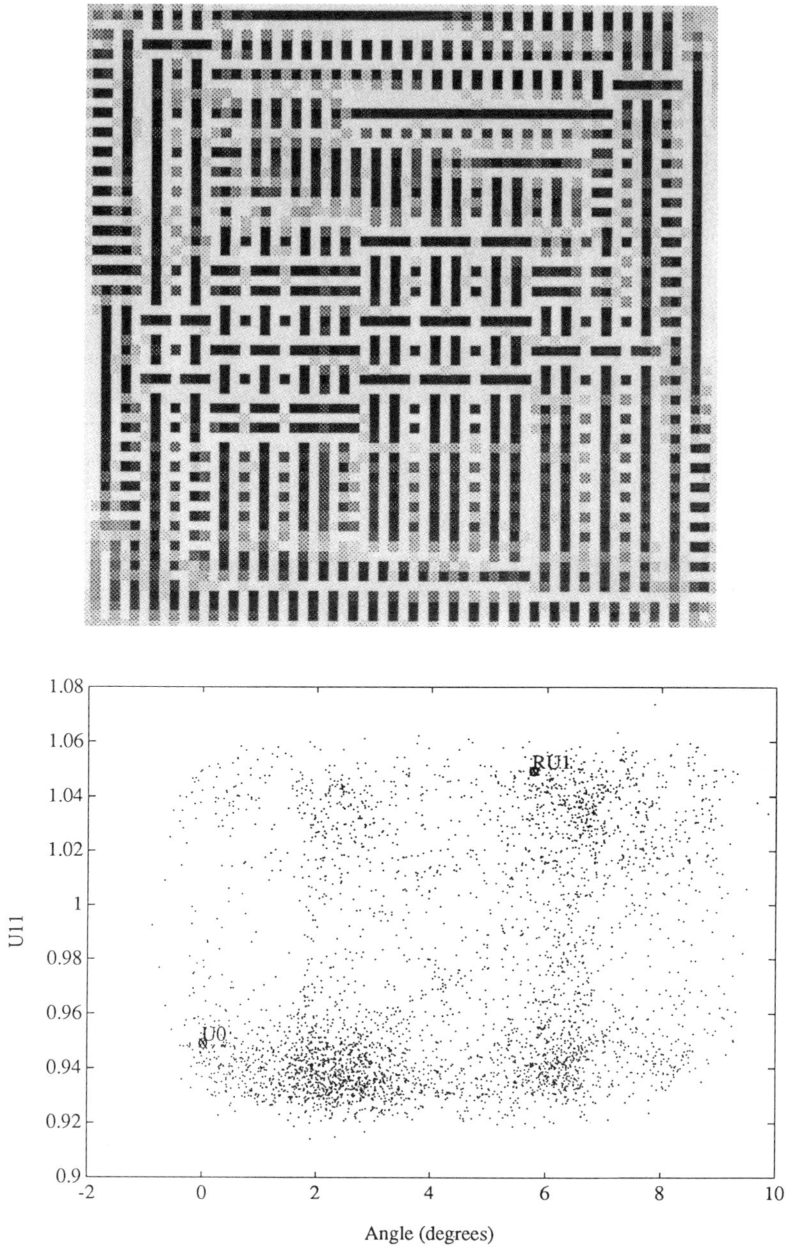

Figure 6. Variant map (top) and gradient distribution (bottom) for 45° problem ($\lambda = 1/2$, $h = 1/64$)

REFERENCES

A F. W. ALT, K.–H. HOFFMANN, M. NIEZGÓDKA AND J. SPREKELS, *A numerical study of structural phase transitions in shape memory alloys*, Preprint #90, Institut für Mathematik, Universitat Augsburg (1985).

BJ1 J. M. BALL AND R. D. JAMES, *Fine phase mixtures as minimizers of energy*, Arch. Rational Mech. Anal., 100 (1987), pp. 13–52.

BJ2 ———, *Experimental tests of a theory of fine microstructure and the two-well problem*, preprint.

Ch M. CHIPOT, *Hyperelasticity for crystals*, European Journal of Applied Math, 1 (1990), pp. 113–129.

CC M. CHIPOT AND C. COLLINS, *Numerical approximations in variational problems with potential wells*, IMA Preprint # 696.

CK M. CHIPOT AND D. KINDERLEHRER, *Equilibrium configurations of crystals*, Arch. Rational Mech. Anal, 103 (1988), pp. 237–277.

C C. COLLINS, *Thesis*, University of Minnesota, 1990.

CKL C. COLLINS, D. KINDERLEHRER, AND M. LUSKIN, *Numerical approximation of the solution of a variational problem with a double well potential*, SIAM J. Numer. Anal. (to appear).

CL1 C. COLLINS AND M. LUSKIN, *The computation of the austenitic-martensitic phase transition*, in *Partial Differential Equations and Continuum Models of Phase Transitions, Lecture Notes in Physics 344, M. Rascle, D.Serre, and M. Slemrod, eds.*, Springer-Verlag, Berlin, New York, 1989, pp. 34–50.

CL2 ———, *Computational results for phase transitions in shape memory materials*, in *Smart Materials, Structures, and Mathematical Issues, C. Rogers, ed.*, Technomic Publishing Co., Lancaster, Pennsylvania, 1989, pp. 198–215.

CL3 ———, *Optimal order error estimates for the finite element approximation of the solution of a nonconvex variational problem*, Math Comp (to appear).

CLR C. COLLINS, M. LUSKIN AND J. RIORDAN, *Computational Images of Crystalline Microstructure*, Video, AMS.

E1 J. L. ERICKSEN, *Some constrained elastic crystals*, in *Material Instabilities in Continuum Mechanics and Related Problems, J. M. Ball, ed.*, Oxford University Press, Oxford, 1987, pp. 119–137.

E2 ———, *Constitutive theory for some constrained elastic crystals*, Int. J. Solids and Structures, 22 (1986), pp. 951–964.

JK R. D. JAMES AND D. KINDERLEHRER, *Theory of diffusionless phase transitions*, in *Partial Differential Equations and Continuum Models of Phase Transitions, Lecture Notes in Physics 344, M. Rascle, D.Serre, and M. Slemrod, eds.*, Springer-Verlag, Berlin, New York, 1989, pp. 51–84.

K D. KINDERLEHRER, *Remarks about equilibrium configurations of crystals*, in *Material Instabilities in Continuum Mechanics and Related Problems, J. M. Ball, ed.*, Oxford University Press, Oxford, 1987, pp. 217–242.

S S. A. SILLING, *Phase changes induced by deformation in isothermal elastic crystals*, J Mech Phys Solids, 37 (1989), pp. 293–316.

COMPUTATIONAL RESULTS FOR A TWO-DIMENSIONAL MODEL OF CRYSTALLINE MICROSTRUCTURE*

CHARLES COLLINS†, MITCHELL LUSKIN‡,
AND JAMES RIORDAN‡.

Abstract. We describe a two-dimensional model of crystalline martensitic microstructure, and we present a new visualization of computational results for the finite element approximation of solutions to the variational problem with microstructure on a sequence of refined meshes.

1. Introduction. We present computational results for a two-dimensional model of crystalline martensitic microstructure which was proposed by Ericksen and James. This two-dimensional model is the reduction of a three-dimensional model for crystals which undergo a cubic to tetragonal phase transformation [BJ2]. It has the property that the energy density is frame-indifferent and has two symmetry-related energy wells. Variational problems of this type do not always attain their minimum value for any admissible deformation [BJ2]. Rather, the solution must often be described in terms of a microstructure since the deformation gradients of minimizing sequences can have oscillations with amplitude which remains finite and with wavelength which converges to zero.

A mathematical theory of microstructure has been developed during the past several years to describe solutions to these variational problems [BJ1, ChKi, E1, E2, J, Ki, Ko]. This theory also gives a recipe for the evaluation of macroscopic properties of crystals with microstructure.

Computations for a three-dimensional model for InTl, a shape-memory alloy with symmetry-related (martensitic) variants, were given in [CoL]. These computations successfully obtained microstructure on the scale of the grid and the austenitic-martensitic interface. We have found that the computation of three-dimensional deformations for crystals with symmetry-related microstructure requires large amounts of computing time. For suitable boundary conditions, the three-dimensional model for InTl has a two-dimensional solution which can be obtained from a two-dimensional model [BJ2]. We have developed the two-dimensional model so that we can more quickly experiment with our algorithms and so that we can more easily do computations which are relevant to the development of the continuum theory. For instance, the two-dimensional model has been used to study complex microstructure involving the mixture of more than two deformation gradients [Co].

*This work is part of the Transitions and Defects in Ordered Materials Project and was supported in part by the National Science Foundation and AFOSR through grants DMS 835-1080 and DMS 871-8881, the ARO through grants DAAL03-88-K-0110 and DAAL03-89-G-0081, the Army High Performance Computing Research Center, the Cray Research Foundation, and by a grant from the Minnesota Supercomputer Institute.

†Department of Mathematics, University of Michigan, Ann Arbor, Michigan 48109.

‡School of Mathematics, University of Minnesota, Minneapolis, Minnesota 55455.

The development of a theory to rigorously analyze the numerical approximation of microstructure in crystals was begun in [CoKiL, CoL] and has been further developed in [ChCo]. These concepts have also recently been applied to the numerical approximation of the fine scale structure of the magnetization field of some ferromagnetic materials [LM].

2. Two-dimensional model. The bulk energy of a two-dimensional crystal is modeled by

$$\mathcal{E}(y) = \int_{\Omega} \phi(\nabla y(x), \theta)\, dx$$

where $\Omega \subset \mathbf{R}^2$ is the reference configuration for the crystal, $y(x) : \Omega \to \mathbf{R}^2$ is the deformation, θ is the temperature, and ϕ is the energy density.

We use the energy density proposed by Ericksen and James which is given by

$$\phi(F, \theta) = \kappa_1(\operatorname{Trace} C - 2)^2 + \kappa_2 c_{12}^2 + \kappa_3 \left(\left(\frac{c_{11} - c_{22}}{2} \right)^2 - \epsilon^2 \right)^2$$

where

$$C = F^T F = \begin{pmatrix} c_{11} & c_{12} \\ c_{21} & c_{22} \end{pmatrix}$$

is the Cauchy-Green strain tensor; κ_1, κ_2, and κ_3 are elastic moduli; and ϵ is the transformation strain. The elastic moduli and the tranformation strain depend on the temperature.

The energy density ϕ is frame-indifferent, i.e.,

$$\phi(RF, \theta) = \phi(F, \theta)$$

for any rotation R and ϕ has the symmetry group of the square, i.e.,

$$\phi(FR_\eta, \theta) = \phi(F, \theta) \qquad \text{for } \eta = \pi/2, \pi, 3\pi/2$$

where R_η is the rotation matrix with angle η

$$R_\eta = \begin{pmatrix} \cos\eta & -\sin\eta \\ \sin\eta & \cos\eta \end{pmatrix}.$$

The energy density $\phi(F, \theta)$ attains its minimum value at the symmetry-related deformation gradients $F^T F = U_0^2$ or $F^T F = U_1^2$ where

$$U_0 = \begin{pmatrix} \sqrt{1-\epsilon} & 0 \\ 0 & \sqrt{1+\epsilon} \end{pmatrix} \qquad \text{and} \qquad U_1 = \begin{pmatrix} \sqrt{1+\epsilon} & 0 \\ 0 & \sqrt{1-\epsilon} \end{pmatrix}.$$

The deformation gradients $F = U_0$ and $F = U_1$ which represent different "variants" of martensite are symmetry-related since

$$U_0 = R_{-\pi/2} U_1 R_{\pi/2}.$$

For an unstressed solid the above model allows the existence of interfaces which separate different variants of martensite. These interfaces or "twin lines" are given by lines across which the deformation is continuous, but across which the deformation gradient is discontinuous. A continuous deformation $y(x)$ exists such that

$$\begin{aligned} \nabla y(x) &= U_0 && \text{where } x \cdot n > 0, \\ \nabla y(x) &= R_\zeta U_1 && \text{where } x \cdot n < 0 \end{aligned}$$

if and only if

$$R_\zeta U_1 = U_0 + a \otimes n \tag{1}$$

for some angle ζ and some vector a. The set of solutions to (1) is given by ζ such that $\cos\zeta = \sqrt{1-\epsilon^2}$, $\sin\zeta = \pm\epsilon$, and

$$a = \sqrt{2}\epsilon\left(\sqrt{1-\epsilon}, \pm\sqrt{1+\epsilon}\right), \qquad n = \frac{\sqrt{2}}{2}(1, \mp 1).$$

Thus, there are two possible families of parallel interfaces.

For $\beta(s)$ taking only the values 0 and 1 the continuous deformation

$$y(x) = U_0 x + a \int_0^{x \cdot n} \beta(t)dt,$$

has discontinuous deformation gradients on lines orthogonal to n since

$$\begin{aligned} \nabla y(x) &= U_0 && \text{where } \beta(x \cdot n) = 0, \\ \nabla y(x) &= R_\zeta U_1 && \text{where } \beta(x \cdot n) = 1. \end{aligned}$$

It follows that $\phi(\nabla y(x)) = 0$ for all $x \in \Omega$.

The unstressed austenitic phase is represented in this model by the deformation gradient $F = I$ where I is the identity matrix, and the unstressed martensitic phase is represented by the symmetry-related deformation gradients $F = U_0$ and $F = U_1$ which represent different "variants" of the martensite. We assume that the temperature θ is held fixed in the body below the transition temperature between the austenitic phase and the martensitic phase. At such a temperature the unstressed martensitic phase should be a global minimum of the energy density while the unstressed austenitic phase should be a local minimum of the energy density.

We recall that for any 2×2 nonsingular matrices V_1 and V_2 such that the eigenvalues of $(V_1 V_2^{-1})^T (V_1 V_2^{-1})$ satisfy $\lambda_1 < 1 < \lambda_2$, there exists a rotation R and nonzero vectors b and m such that [BJ1]

$$RV_1 = V_2 + b \otimes m.$$

In particular, there exists a rotation R and nonzero vectors b and m such that

$$RI = U_0 + b \otimes m. \tag{2}$$

Now (2) implies that the two-dimensional model allows a continuous deformation with an interface separating a region of austenite $F = RI$ from a region of martensite $F = U_0$. However, the three-dimensional model of InTl does not allow an interface to separate a region of austenite from a region containing a single variant of martensite, and this is confirmed by experimental observations. Rather, for the three-dimensional model an interface can separate regions of austenite and martensite only if the martensitic region is a fine-scale mixture of more than one variant of martensite [BJ1]; and this is also confirmed by experimental results. By modifying the two-dimensional energy density so that $F = I$ is not a local minimum, we have eliminated the possibility of a spurious interface between regions of austenite and of single-variant martensite for a two-dimensional unstressed crystal.

We chose the values of the material coefficients to resemble the elastic moduli and transformation strain for InTl which has a cubic austenitic phase and a tetragonal martensitic phase. To simulate the properties of the three-dimensional energy density proposed by Ericksen for InTl which has a cubic symmetry group [E1, E2], we have constructed the two-dimensional energy density ϕ with the symmetry group of the square and we have used the moduli

$$\kappa_1 = 10, \quad \kappa_2 = 3, \quad \kappa_3 = 1, \quad \epsilon = .1.$$

The transformation strain ϵ is taken to be larger than the physical strain to enable numerical computations to be done on a coarser grid. We have also represented the tetragonal martensitic phase by a two-dimensional rectangular phase.

The goal of our computations is to compute the displacement, y_h, which minimizes the bulk energy among all admissible finite element displacements on a mesh with length scale h. The deformation on the boundary of the body is constrained to equal

$$y(x) = \left(\frac{1}{2}U_0 + \frac{1}{2}R_\zeta U_1\right)x, \qquad x \in \partial\Omega. \tag{3}$$

There does not exist a deformation which has the minimum energy $\mathcal{E} = 0$ and which satisfies the boundary conditions (3) [BJ2]. Rather, if $\mathcal{E}(y_n) \to 0$ as $n \to \infty$ for a sequence of deformations satisfying the boundary conditions, then the amplitude of the oscillations of $\nabla y_n(x)$ remains finite as $n \to \infty$, but the wavelength of the oscillations becomes arbitrarily small. The "microstructure" solution to this problem is unique, though, and is given by the mixture of U_0 and $R_\zeta U_1$ in equal proportions [BJ2].

We can give an analytic treatment of the microstructure for this problem by defining

$$\beta(x) = \begin{cases} 0 & \text{if } 2m < x < 2m+1, \text{ for } m \text{ an integer} \\ 1 & \text{if } 2m-1 < x < 2m, \text{ for } m \text{ an integer} \end{cases}$$

and

$$\beta_\delta(x) = \beta(x/\delta).$$

Then

$$y_\delta(x) = U_0 x + a \int_0^{x \cdot n} \beta_\delta(t)dt$$

has minimum energy ($\mathcal{E}(y_\delta) = 0$), but $y_\delta(x)$ does not satisfy the boundary conditions (3). However,

$$y_\delta(x) \to \left(\frac{1}{2} U_0 + \frac{1}{2} R_\zeta U_1 \right) x$$

uniformly as $\delta \to 0$. It follows that we can modify $y_\delta(x)$ near the boundary to construct a deformation, $\tilde{y}_\delta(x)$, which satisfies the boundary conditions (3) and such that

$$\mathcal{E}(\tilde{y}_\delta) \to 0 \qquad \text{as } \delta \to 0.$$

The scale of the microstructure for $\tilde{y}_\delta(x)$ is δ.

3. Two-dimensional computations. We used continuous, bilinear finite elements, and we developed an optimization algorithm based on the conjugate gradient algorithm [Co]. The crystal was oriented so that the lines of discontinuity of the deformation gradient are diagonal to the mesh—the most difficult test.

In each local element, we evaluate the deformation gradient of y_h at the center, and we color the area in the local element to display the function

$$\psi(F) = \frac{\|F^T F - U_0^2\|}{\|F^T F - U_0^2\| + \|F^T F - U_1^2\|}.$$

In Figure 2 (see color insert that follows this chapter) we see the results of our numerical experiments with the boundary conditions (3) where $\sin \zeta = -\epsilon$ for mesh lengths $h = 1/16$, $h = 1/32$, $h = 1/48$, and $h = 1/64$. The color representation of ψ is shown in Figure 1. The element is colored purple when $F^T F \approx U_0^2$, the element is colored yellow when $F^T F \approx U_1^2$, and the element is colored blue or green when F is not in either of the energy wells.

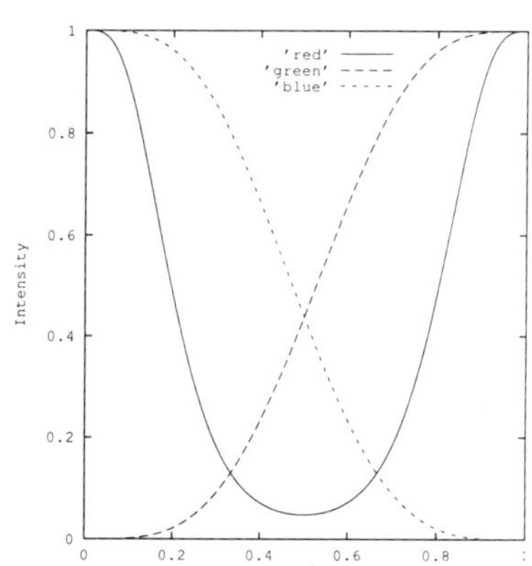

Figure 1. Color representation of ψ.

Our algorithm has successfully computed a microstructure on the scale of each successively finer mesh. Since the computed microstructure is not completely regular, we have actually computed a local minimum of the finite element optimization problem. Nevertheless, our computed local minimum has small enough energy so that it exhibits the microstructure of the global minimum.

The number of local minima becomes arbitrarily large as the mesh is refined since we can have local minima with oscillations of arbitrarily small wavelength and since the number of possible "defects" where the microstructure is irregular can become arbitrarily large.

The video [CoLR] shows the path of our conjugate gradient algorithm to a local minimum. The microstructure organizes itself so that the energy density is small in disjoint regions. As these regions coalesce or approach the boundary, the unique microstructure that is compatible with the boundary conditions is chosen throughout the entire crystal.

REFERENCES

[BJ1] J. M. Ball and R. D. James, *Fine phase mixtures as minimizers of energy*, Arch. Rational Mech. Anal., 100 (1987), pp. 13–52.

[BJ2] J. M. Ball and R. D. James, *Experimental tests of a theory of fine microstruture and the two-well problem*, preprint.

[ChCo] M. Chipot and C. Collins, *Numerical approximations in variational problems with potential wells*, IMA Preprint 696.

[ChKi] M. Chipot and D. Kinderlehrer, *Equilibrium configurations of crystals*, Arch. Rational Mech. Anal, 103 (1988), pp. 237–277.

[Co] C. Collins, *Computation of twinning*, this volume.

[CoKiL] C. Collins, D. Kinderlehrer, and M. Luskin, *Numerical approximation of the solution of a variational problem with a double well potential*, SIAM J. Numer. Anal., 28 (1991), pp. 321–332.

[CoL1] C. Collins and M. Luskin, *The computation of the austenitic-martensitic phase transition*, in *Partial Differential Equations and Continuum Models of Phase Transitions, Lecture Notes in Physics 344, M. Rascle, D.Serre, and M. Slemrod, eds.*, Springer-Verlag, 1989, pp. 34–50.

[CoL2] C. Collins and M. Luskin, *Optimal order estimates for the numerical approximation of the solution of a variational problem with a double well potential*, to appear, Math. Comp. (to appear).

[CoLR] C. Collins, M. Luskin, and J. Riordan, *Computational images of crystalline microstructure*, in *AMS Video Series*, 1991.

[E1] J. L. Ericksen, *Some constrained elastic crystals*, in *Material Instabilities in Continuum Mechanics and Related Problems, J. M. Ball, ed.*, Oxford University Press, 1987, pp. 119–137.

[E2] J. L. Ericksen, *Constitutive theory for some constrained elastic crystals*, Int. J. Solids and Structures, 22 (1986), pp. 951–964.

[J] R. James, *Basic principles for the improvement of shape-memory and related materials*, in *Smart Materials, Structures, and Mathematical Issues, C. Rogers, ed.*, Technomic Publishing Co., 1989.

[Ki] D. Kinderlehrer, *Remarks about equilibrium configurations of crystals*, in *Material Instabilities in Continuum Mechanics and Related Problems, J. M. Ball, ed.*, Oxford University Press, 1987, pp. 217–242.

[Ko] R. Kohn, *Relaxation of a double-well energy*, Continuum Mechanics and Thermodynamics (to appear, 1991).

[L] M. Luskin, *Numerical analysis of microstructure for crystals with a nonconvex energy density*, in *in The Metz Days Surveys, 1989-90*, Pitman Research Notes in Mathematics, M. Chipot and J. Saint Jean Paulin, eds.,, Longman Company, UK (to appear).

[LM] M. Luskin and L. Ma, *Analysis of the finite element approximation of microstructure in micromagnetics*, SIAM J. Numer. Anal. (to appear).

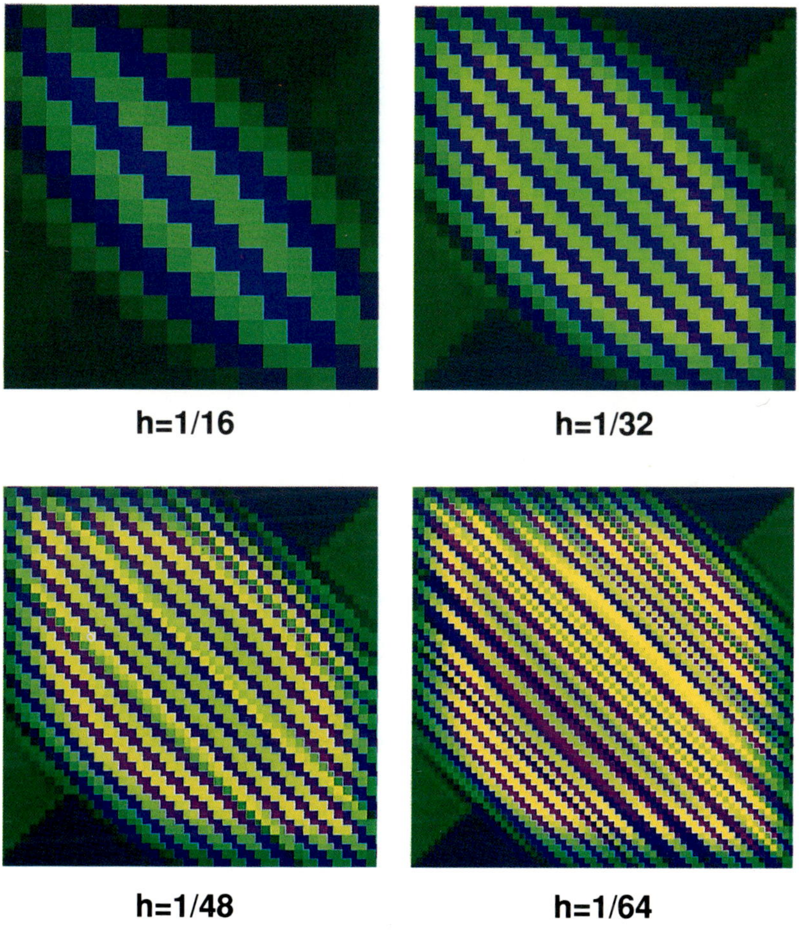

FIGURE 2 *(Collins, Luskin, Riordan)*. Deformation gradient of $yh(x)$ for meshes with scales h = 1/16, h = 1/32, h = 1/48, and h = 1/64. The deformation gradient at the midpoint of each local element is displayed. The local element is colored according to the value of $\psi(F)$ and the representation of Figure 1.

LOCAL BIFURCATION THEORY FOR THERMOELASTIC BRAVAIS LATTICES

J.L. ERICKSEN*

Abstract. Some weak first order phase transitions encountered in shape-memory alloys seem to occur near bifurcations which are not actually observed. Being curious about the nature of these in particular, I began to think about the nature of all the bifurcation patterns predicted by the theory of thermoelastic Bravais lattices. Presented here are local characterizations of the most likely patterns near a bifurcation.

1. Introduction. As is made evident by some of the papers collected in reference [1], for example, nonlinear thermoelasticity theory describing Bravais lattices is proving to be useful for analyzing near-transition behavior of alloys exhibiting shape-memory effects. These involve Martensitic transformations, commonly of first order. From observations such as are discussed in various places in reference [2], it seems typical of these transformations that, as one cools an unloaded crystal which is in the Austenitic phase at higher temperatures, some linear elastic modulus becomes unusually small, seeming to extrapolate to zero at a temperature slightly below that at which the crystal transforms to the Martensitic phase. While there is little experimental evidence to support it, there is some opinion that a similar effect is likely to occur as the Martensitic phase is warmed, to get near the transformation temperature.

Theoretical experience suggests that the vanishing of such moduli will be associated with some kind of bifurcation, probably involving equilibrium branches too unstable to be observable. This led me to be curious as to what might be deduced from theory about the natures of such bifurcation patterns. Briefly and roughly, as I interpret it, the general problem is to get a qualitative description of the bifurcation patterns most likely to occur in thermoelastic Bravais lattices. There is a global problem, involving one or more connected sets of equilibrium branches, with any number of bifurcations. For the most part, I'll be concerned only with the nature of bifurcation patterns near one bifurcation, the local problem. A few remarks about a more global problem are included in Section 10.

On the face of it, this seems to involve a large number of possibilities. Physically, equilibria can have any of the symmetries associated with the seven point groups which can be realized in Bravais lattices, which are described in the Appendix, and branches with different symmetries can meet at bifurcations. With each point group is associated some number of elastic moduli, giving different possible choices of that which vanishes at bifurcation. However, we will see that there are ways to reduce these to just six different mathematical problems which I found tractable, although I am not expert in bifurcation theory, although I have not yet found references to

*Department of Aerospace Engineering and Mechanics and School of Mathematics, University of Minnesota, Minneapolis, MN 55455. Member of the Research Group on Transitions and Defects in Ordered Materials.

all. However, I found it hard to make some points without giving some indication of how the analysis can be done, so I have done this. Mathematically, there are additional possibilities but, by an old criterion which is to be discussed and used, they are unlikely to be encountered.

My purpose is to elaborate these remarks. To keep this paper from being unreasonably long, I will omit some of the details in various analyses, but indicate the kind of reasoning which is used in getting the results.

There are some different aspects to this study. One is the reduction of the numerous possibilities to the aforementioned six problems. One needs to understand this in some detail, to know how to relate any particular physical possibility to one of the six mathematical problems. then, one needs some analysis of each of the six and, as noted above, I'll sketch mine. Finally, I'll discuss a particular case which seems to represent a rather common occurrence, experimentally.

Some of the elementary bifurcation theory for Bravais lattice is discussed by Ericksen [3], but the present study goes far beyond this, to give a fairly complete local theory of the generic kind.

2. Basic formulation. Here, the absolute temperature is considered to be uniform and to be a control variable, and we will consider only homogeneous configurations of our Bravais lattices. In our analyses, some homogeneous configuration will be taken as a reference so, for the configurations considered, the right Cauchy–Green tensor $\mathbf{C}$ will be independent of position. According to thermoelasticity theory, the Helmholtz free energy density ϕ is a function of the form

$$\phi = \widehat{\phi}(\mathbf{C}, \theta) \ . \tag{21}$$

While there are physical reasons to be interested in cases where $\widehat{\phi}$ exhibits rather mild singularities at bifurcations, I think it better to avoid these complications, for a first look. At least I want to avoid the rather common assumption of analyticity, and its implications relating to analytic continuation. Thus, in the neighborhoods considered, I'll assume more than is necessary, that $\widehat{\phi}$ is a $\mathbf{C}^{\infty}$ function: different weaker assumptions suffice for different analyses.

As is discussed by Ericksen [14], for example, versions of such theory appropriate for Bravais lattices involve some special features. Briefly, $\widehat{\phi}$ is to be invariant under an infinite discrete group conjugate to $GL(3, \mathbb{Z})$, the group of general linear transformations on the integers, in three dimensions. However, the restriction of $\widehat{\phi}$ to a suitably chosen configuration, with $\mathbf{C} = \mathbf{C}_0$, say is invariant only under a finite subgroup, that which leaves $\mathbf{C}_0$ invariant. If one chooses this to be the reference so that $\mathbf{C}_0 = 1$, the subgroup is just the point group P_0 associated with this particular configuration, and all other configurations in the neighborhood have point groups which are subgroups of it. If such a configuration is associated with a value $\mathbf{C} = \overline{\mathbf{C}}$, it will have a point $\overline{P}$ described as follows:

$$\overline{P} = \{\mathbf{Q} \in P_0 \mid \mathbf{Q}^T \overline{\mathbf{C}} \mathbf{Q} = \overline{\mathbf{C}}\} \tag{2.2}$$

it being understood that, with $\mathbf{C}_0 = 1$, any possible $\mathbf{Q}$ is an orthogonal transformation, one of those described in the Appendix.

For us, equilibria will mean unstressed configurations, values of $\mathbf{C}$ and θ satisfying

$$\frac{\partial \phi}{\partial \mathbf{C}} = 0 , \tag{2.3}$$

If $(\mathbf{C}_0,\ \theta_0)$ is one, one can try to use the implicit function theorem to infer the existence of smooth equilibrium paths $\mathbf{C} = \widehat{\mathbf{C}}(\theta)$ through it, on which (2.3) is satisfied. What is important for this are some properties of the fourth order tensor

$$\mathcal{B}(\mathbf{C}_0, \theta_0) = \frac{\partial^2 \widehat{\phi}}{\partial \mathbf{C} \partial \mathbf{C}} (\mathbf{C}_0,\ \theta_0) \tag{2.4}$$

or, what is equivalent, $\mathfrak{a}$, the tensor of linear elastic moduli. Since properties of the latter are more likely to be familiar to the reader, I'll use it. To get $\mathfrak{a}$ from $\mathcal{B}$, one changes the reference configuration to make $\mathbf{C}_0 = \mathbf{1}$, then takes

$$\mathfrak{a} = 4\mathcal{B}(\mathbf{1}, \theta_0) \ . \tag{2.5}$$

In a way, this is a bit awkward since, if one is evaluating $\mathfrak{a}$ on a smooth branch, $\mathbf{C} = \widehat{\mathbf{C}}(\theta)$, one needs to continually change the reference configuration to make $\widehat{\mathbf{C}}(\theta) = \mathbf{1}$, but this is what fits rather common practice.

As defined, $\mathfrak{a}$ can be interpreted in the usual way, as a symmetric linear transformation on the space of symmetric second-order tensors. This involves considering the latter as equipped with the inner product.

$$\mathbf{A} \bullet \mathbf{B} = \operatorname{tr} \mathbf{AB} \ , \tag{2.6}$$

with $\mathbf{A}$ and $\mathbf{B}$ representing any pair of symmetric second order tensors. In this setting, $\mathfrak{a}$ then has real eigenspaces with the usual orthogonality properties.

Now, from the view of the implicit function theorem, the trivial case occurs when

$$\ker \mathfrak{a} = \emptyset \ , \tag{2.7}$$

in which case one has the local existence and uniqueness of the aforementioned smooth branch. On such a branch, as long as (2.7) holds, it follows that the point group is the same as that at the (fixed) reference, P_0, also that the values of $\widehat{\mathbf{C}}(\theta)$ are restricted by the obvious interpretation of (2.2): one can read off the possible forms from the description by Truesdell [5, Ch. IX], for example. Also, $\mathfrak{a}$, evaluated at any point on such a branch, will be invariant under P_0. So, these are the normal branches, commonly considered in the linear thermoelasticity theory, for example. In fact, such theories also presume that the eigenvalues of $\mathfrak{a}$ are positive.

Our interest is more in limit points of such branches, at which (2.7) fails to hold. By continuity, they will satisfy (2.3), and the point group at the limit point will either be or include as a subgroup that on any of the normal branches having it as a limit point. Said differently, these are the places where bifurcations can occur.

Now, the intent is not to try to find all of the mathematical possibilities, but to look for the most likely. To assess this, I'll use an old criterion, still commonly used by physicist, which I trust. One could use instead more modern versions of generic bifurcation theory, but this involves making some sticky decisions about what is the appropriate topology, *etc.*, like those discussed by Man [6] in his discussion of Gibbs' phase rule. In any event, what I'll do is to consider things of interest, like $\mathfrak{a}$, evaluated on a normal branch, as function of θ. As a consequence of invariance, they will satisfy some equations, to be viewed as identities. Otherwise, the view is that one equation might well be satisfied at a particular value of θ, but it is unlikely that two or more independent equations will there be satisfied. There is a physical idea which supplies some motivation for accepting such ideas, which is also covered in the interesting discussion by Man [6]. Briefly, it involves the idea that a material is described not by one constitutive equation, but a collection of slightly different equations, an idea which appeals to me.

In particular, with a value of $\mathfrak{a}$ being invariant under one of the seven point groups, its eigenspaces will also have this property and, in some cases, this require that some of the eigenvalues be equal. By considering spectral representations, one can show that, when the number of distinct eigenvalues is as large as it can be, the eigenspaces are irreducible. That is, they contain no proper subspaces which are invariant under the applicable point group. It might well happen that two which could be different do coincide at a particular value of θ, on a branch, but this would count as an equation. Now, for $\mathfrak{a}$ to have a non-trivial kernel, one of its eigenvalues must vanish. By our criterion, it is then unlikely that two eigenvalues will be equal unless, as a consequence of invariance, the two must be equal. This provides us with one important conclusion, *viz.*.

(2.8)

Most likely, when $\ker \mathfrak{a} \neq \emptyset$, *the eigenspaces of* $\mathfrak{a}$, *in particular* $\ker \mathfrak{a}$, *are irreducible* .

In Section 10, I illustrate a subtlety associated with this. Briefly, it is not considered unlikely that the point group at a limit point of some branch might be larger than that on the branch, and, when this happens, it might require some coalescence of eigenvalues or other conditions not then to be considered as equations. Later, we will make use of some other consequences of our criterion, but this one will be used repeatedly. So, for any of the seven point groups, an eigenvalue associated with any of the irreducible subspaces might vanish, and we need to consider all such possibilities. One could envisage the possibility that the list might be enlarged, by some consideration of the nature of branches for which the point considered is a limit point but, as we shall see, this is a problem which takes care of itself.

3. General scheme for reducing problems. Essentially, we follow common practice in bifurcation theory, with some accounting for the quirks induced by invariance. The reference configuration is considered to be fixed, at a place where $\ker \mathfrak{a} \neq \emptyset$, at some particular temperature θ_0. So, by assumption

$$\ker \mathfrak{a} \neq \emptyset \tag{3.1}$$

and, is suggested by (2.8), we also assume that the eigenspaces of $\mathfrak{a}$ are irreducible. Then, $\ker \mathfrak{a}$ is an irreducible space of some dimension N. Later, it will become clear that $N \leq 3$. As a basis for symmetric second-order tensors, we introduce an orthonormal basis of eigenvectors of $\mathfrak{a}$, denoted by $\mathbf{E}_i$ $(i = 1, \ldots, 6)$,

(3.2) $$\mathbf{E}_i \bullet \mathbf{E}_j = \delta_{ij} \; (i,\; j = 1, \ldots, 6) \; ,$$

with $\mathbf{E}_1, \ldots, \mathbf{E}_N$ spanning $\ker \mathfrak{a}$. In particular, values of $\mathbf{C}$ are represented by components x_i as indicated by

(3.3) $$\mathbf{C} = \mathbf{1} + \sum_{i=1}^{6} x_i \mathbf{E}_i \; ,$$

with $x_i = 0$ at the putative bifurcation point. By the implicit function theorem, we can solve (2.3) locally to get relations of the form

(3.4) $$x_j = f_j \; (x_1, \ldots, x_N, \theta), j = N+1, \ldots, 6$$

these functions being unique and smooth: they also satisfy the conditions

(3.5) $$f_j(0, \ldots, 0, \theta_0) = 0 \; .$$

Substituting these into the function $\widehat{\phi}$, we get the reduced potential, represented by a smooth function of the form

(3.6) $$\mathbf{\Phi}(x_1, \ldots, x_N, \theta) \; .$$

Then, by standard reasoning, the remaining equilibrium equations reduce to

(3.7) $$\frac{\partial \mathbf{\Phi}}{\partial x_i} = 0$$

and

(3.8) $$\frac{\partial \mathbf{\Phi}}{\partial x_i} = \frac{\partial^2 \mathbf{\Phi}}{\partial x_i \partial x_j} = 0 \quad \text{at} \quad x_i = 0, \; \theta = \theta_0 \; ,$$

where the indices take on the obvious N values. As I interpret the jargon used by physicists, $x_1 \ldots, x_N$ are "order parameters."

From remarks made earlier, $\widehat{\phi}$ can be considered to be invariant under the point group P_0 associated with the reference configuration and, from this and the aforementioned uniqueness of solutions, $\mathbf{\Phi}$ inherits some invariance, which needs to be determined. The relevant point group involves some orthogonal transformations $\mathbf{Q}$, noted in the Appendix, transforming $\mathbf{C}$ to

(3.9) $$\mathbf{Q}^T \mathbf{C} \mathbf{Q} = \mathbf{1} + \sum_{i=1}^{6} x_i \; (\mathbf{Q}^T \mathbf{E}_i \mathbf{Q}) \; ,$$

and with $\mathbf{E}_i$ as a basis, we can determine numbers λ_i^j such that

$$\mathbf{Q}^T\mathbf{E}_i\mathbf{Q} = \sum_{j=1}^{6} \lambda_i^j \mathbf{E}_i \ . \tag{3.10}$$

Of course, this induces on x_i the transformation to the components $\overline{x}_i$ of $\mathbf{Q}^T\mathbf{C}\mathbf{Q}$ given by

$$\overline{x}_i = \sum_{i=1}^{6} \lambda_i^j x_j, \quad i,j = 1,\ldots,6 \ . \tag{3.11}$$

Now, since the inner product (2.6) is invariant under all orthogonal transformations, the $\mathbf{Q}^T\mathbf{E}_i\mathbf{Q}$ also form an orthonormal basis, whence follows that

$$\Lambda = \|\lambda_k^j\| \ , \quad i,j = 1,\ldots,6 \tag{3.12}$$

is an orthogonal matrix, of a rather special kind. That is, since the eigenspaces of $\mathfrak{a}$ are invariant under P_0, Λ must map any such eigenspace onto itself. Thus, the obvious restriction of this linear transformation to an eigenspace is described by a matrix which is also orthogonal. In particular, this is true of the submatrix

$$\lambda = \|\lambda_i^j\| \ , \quad i,j = 1,\ldots,N \ . \tag{3.13}$$

describing the transformation of order parameters, so it satisfies

$$\lambda^{-1} = \lambda^T \ . \tag{3.14}$$

Also, it is easy to show that the matrices thus obtained from a group under matrix multiplication, which I'll call $\mathfrak{R}(P_0)$ the reduction of P_0,

$$\lambda \in \mathfrak{R}(P_0) \ , \tag{3.15}$$

Bear in mind that it depends on which eigenspace is considered to be $\ker \mathfrak{a}$, as well as P_0. To simplify notation a bit, we now write

$$\mathbf{y} = (x_1,\ldots,x_N) \ . \tag{3.16}$$

After a bit of calculation, using the uniqueness of the solutions (3.4), one finds that $\mathfrak{R}(P_0)$ is the invariance group for the reduced potential, so

$$\mathbf{\Phi}(\lambda\mathbf{y},\theta) = \mathbf{\Phi}\ (\mathbf{y},\theta) \ , \quad \lambda \in \mathfrak{R}(P_0) \ , \tag{3.17}$$

at least for $\mathbf{y}$ near the origin. Here, I have glossed one point, in that (3.17) makes no sense unless the domain of $\mathbf{\Phi}$ is invariant under $\mathfrak{R}(P_0)$. One can take care of this by considering the domain to be a suitably chosen neighborhood.

Another item of interest is the point group associated with $\mathbf{y}_0$, a particular value of $\mathbf{y}$ near the origin. To determine this, one calculates the subgroup $\mathfrak{R}(P_0, \mathbf{y}_0)$ leaving $\mathbf{y}_0$ invariant:

$$\mathfrak{R}(P_0, \mathbf{y}_0) = \{\lambda \in \mathfrak{R}(P_0) \mid \lambda \mathbf{y}_0 = \mathbf{y}_0\} \ . \tag{3.18}$$

With (3.4), $\mathbf{y}_0$ determines a point $\mathbf{x}_0$ in the six-dimensional space $(x_1, \ldots, x_6)$, and λ is a submatrix of at least one 6×6 matrix $\wedge(\lambda)$. From the uniqueness of solutions of (3.4), it then follows that

$$\wedge(\lambda)\mathbf{x}_0 = \mathbf{x}_0 \ . \tag{3.19}$$

In turn, $\wedge(\lambda)$ is associated with at least one $\mathbf{Q} \in P_0$, by (3.11); call it $\mathbf{Q}(\lambda)$. From (3.3), $\mathbf{x}_0$ determines some value $\mathbf{C}_0$ of $\mathbf{C}$ and, with (3.19), we have

$$\mathbf{Q}(\lambda)^T \mathbf{C}_0 \mathbf{Q}(\lambda) = \mathbf{C}_0 \ . \tag{3.20}$$

The linear transformations $\mathbf{Q}(\lambda)$ thus obtained then form some subgroup of P_0 which, as is indicated in (2.2), is the point group associated with this particular configuration. From such theory it follows that the order of this group can be calculated by dividing the order of P_0 by the number of distinct points in the orbit of $\mathbf{C}_0$ under P_0 or, what is the same, the number of distinct points in the orbit of $\mathbf{y}_0$ under $\mathfrak{R}(P_0)$. From the Appendix, it is clear that the order uniquely determines the point group. Frequently, I will use this idea to identify point groups.

Concerning the function $\widehat{\phi}$, we have assumed that it is smooth, has some invariance locally, and admits some equilibria compatible with bifurcation. What is left rather tacit above is the notion that, at $\theta = \theta_0$, $\mathbf{y} = 0$ represents a limit point of one or more normal branches: as it will turn out, it isn't important to make this more explicit. Otherwise, one can show that, for our local analyses, $\mathbf{\Phi}$ need not satisfy any conditions beyond those mentioned explicitly above. Of course, any additional assumptions made about $\widehat{\phi}$ could lead to additional restriction on $\mathbf{\Phi}$, not accounted for here.

To proceed with analysis of possible bifurcations, one needs to know the possible values of N and the groups $\mathfrak{R}(P_0)$. Different physical problems can and do lead to the same N and $\mathfrak{R}(P_0)$ although, in some cases, one needs to make some judicious choices of the eigenvectors, to make such identifications. So, this is the next step to be taken, to construct guide maps to link each of the physical possibilities to one of the mathematical bifurcation problems.

4. Guide maps. Essentially, this section is a tabulation of important information for each of the seven point groups described in the Appendix. It includes characterizations of the possible orthogonal irreducible subspaces which are, for our purposes, the possible eigenspaces. Also, where needed, I'll describe my special choice of the eigenvectors $\mathbf{E}_1, \ldots, \mathbf{E}_N$, one which can be used to get the particular representations of $\mathfrak{R}(P_0)$ noted below.

I'll not discuss derivations in detail, but some general comments might be helpful. One can start with the second-order tensors which are left invariant under the group P_0 which are described by Truesdell [5, Chapter IX], for example. These form a linear space of some dimension. Associated with this are one-dimensional irreducible spaces, equal in number to the aforementioned dimension. If the number exceeds one, the particular form of such eigenvectors can be expected to be different for different materials, *etc.*. for given point group. This space will always include the identity, so all tensors in the orthogonal complement will be traceless. In this complement, there can be some one-dimensional irreducible spaces. If so, they will consist of tensors transforming to themselves by some elements of the group, to their negatives by other, making it easy to locate these.

If $\ker \mathfrak{a}$ is of the first kind, one has

$$N = 1 \ , \quad \mathfrak{R}(P_0) = SO(1) \ , \tag{4.1}$$

where $SO(1)$ is the proper orthogonal group in one dimension, the trivial group consisting only of the identity

$$SO(1) = 1 \ . \tag{4.2}$$

If $\ker \mathfrak{a}$ is of the second kind,

$$N = 1 \ , \quad \mathfrak{R}(P_0) = O(1) \ , \tag{4.3}$$

where $O(1)$, the one-dimensional orthogonal group, also includes reflections

$$O(1) = 1 \ , \quad y_1 \to -y_1 \ . \tag{4.4}$$

Having determined all of these, and the space which they span, one can look in the remaining orthogonal complement for two-dimensional irreducible subspaces. It turns out the possibility that one of these is $\ker \mathfrak{a}$ leads to cases of the following kind

$$N = 2 \ , \quad \mathfrak{R}(P_0) = G(\psi) \ , \quad \psi = \pi/2 \ , \ \pi/3 \text{ or } 2\pi/3 \ . \tag{4.5}$$

where $G(\psi)$ is the group generated by the two matrices

$$\begin{pmatrix} -1 & 0 \\ 0 & 1 \end{pmatrix} , \begin{pmatrix} \cos\psi & \sin\psi \\ -\sin\psi & \cos\psi \end{pmatrix} . \tag{4.6}$$

Having determined the two-dimensional possibilities, one can again take the orthogonal complement, and look for the three-dimensional possibilities. It turns out that there is just one of these, associated with cubic symmetry, and none of higher dimension. For it,

$$\mathfrak{R}(P_0) = G(3) \ , \tag{4.7}$$

where $G(3)$ is a group of order 24, generated by two improper orthogonal transformations $\mathbf{Q}_1$ and $\mathbf{Q}_2$ which, in notation used in the Appendix, can be taken as

$$\mathbf{Q}_1 = -\mathbf{R}(\pi/2,\ \mathbf{e}_1) \quad \text{and} \quad \mathbf{Q}_2 = -\mathbf{R}(\pi/2,\ \mathbf{e}_2)\ . \tag{4.8}$$

With the point groups numbered as in the Appendix, I now tabulate important details. Tensors are described in terms of components, relative to the bases there used.

No. 1

The irreducible subspaces are all one dimensional, and consist of tensors left invariant under the group, orthogonal to each other, so (4.1) *applies to any of them.*

No. 2

Cases to which (4.1) *applies involve four orthogonal one-dimensional spaces spanning the space of tensors of the form*

$$\begin{Vmatrix} a & b & 0 \\ b & c & 0 \\ 0 & 0 & d \end{Vmatrix}$$

Here and below a, b etc *are arbitrary numbers. The remaining irreducible spaces are also one-dimensional, consisting of two orthogonal spaces, spanning the space indicated by*

$$\begin{Vmatrix} 0 & 0 & a \\ 0 & 0 & b \\ a & b & 0 \end{Vmatrix}$$

these being possibilities to which (4.3) *applies.*

No. 3

Cases to which (4.1) *applies involve three one-dimensional spaces, with span of the form*

$$\begin{Vmatrix} a & 0 & 0 \\ 0 & b & 0 \\ 0 & 0 & c \end{Vmatrix}$$

The remaining irreducible spaces are also one-dimensional, those of the form

$$\begin{Vmatrix} 0 & a & 0 \\ a & 0 & 0 \\ 0 & 0 & 0 \end{Vmatrix}, \quad \begin{Vmatrix} 0 & 0 & a \\ 0 & 0 & 0 \\ a & 0 & 0 \end{Vmatrix}$$

and

$$\begin{Vmatrix} 0 & 0 & 0 \\ 0 & 0 & a \\ 0 & a & 0 \end{Vmatrix},$$

with (4.3) *applying to these.*

No. 4

Cases to which (4.1) *applies involve two one-dimensional spaces, with span of the form*

$$\begin{Vmatrix} a & 0 & 0 \\ 0 & a & 0 \\ 0 & 0 & b \end{Vmatrix}.$$

The remaining irreducible spaces consist of two two-dimensional spaces. One is of the form

$$\begin{Vmatrix} a & b & 0 \\ b & -a & 0 \\ 0 & 0 & 0 \end{Vmatrix}.$$

If in it one takes as a basis

$$\sqrt{2}\,\mathbf{E}_1 = \begin{Vmatrix} 0 & 1 & 0 \\ 1 & 0 & 0 \\ 0 & 0 & 0 \end{Vmatrix}, \quad \sqrt{2}\,\mathbf{E}_2 = \begin{Vmatrix} 1 & 0 & 0 \\ 0 & -1 & 0 \\ 0 & 0 & 0 \end{Vmatrix},$$

(4.5) *applies with* $\psi = 2\pi/3$. *The other is of the form*

$$\begin{Vmatrix} 0 & 0 & a \\ 0 & 0 & b \\ a & b & 0 \end{Vmatrix}$$

and, with the basis

$$\sqrt{2}\,\mathbf{E}_1 = \begin{Vmatrix} 0 & 0 & 1 \\ 0 & 0 & 0 \\ 1 & 0 & 0 \end{Vmatrix}, \quad \sqrt{2}\,\mathbf{E}_2 = \begin{Vmatrix} 0 & 0 & 0 \\ 0 & 0 & 1 \\ 0 & 1 & 0 \end{Vmatrix}$$

(4.5) *applies, also with* $\psi = 2\pi/3$.

No. 5

Cases to which (4.1) *applies involve two one-dimensional spaces, with span of the form*

$$\left\|\begin{matrix} a & 0 & 0 \\ 0 & a & 0 \\ 0 & 0 & b \end{matrix}\right\| .$$

There are also two to which (4.3) *applies, one of the form*

$$\left\|\begin{matrix} 0 & a & 0 \\ a & 0 & 0 \\ 0 & 0 & 0 \end{matrix}\right\| ,$$

the other of the form

$$\left\|\begin{matrix} a & 0 & 0 \\ 0 & -a & 0 \\ 0 & 0 & 0 \end{matrix}\right\| .$$

Finally, there is one with dimension two, of the form

$$\left\|\begin{matrix} 0 & 0 & a \\ 0 & 0 & b \\ a & b & 0 \end{matrix}\right\| .$$

With the basis

$$\sqrt{2}\,\mathbf{E}_1 = \left\|\begin{matrix} 0 & 0 & 1 \\ 0 & 0 & 0 \\ 1 & 0 & 0 \end{matrix}\right\| , \quad \sqrt{2}\,\mathbf{E}_2 = \left\|\begin{matrix} 0 & 0 & 0 \\ 0 & 0 & 1 \\ 0 & 1 & 0 \end{matrix}\right\| ,$$

(4.5) *applies, with* $\psi = \pi/2$.

No. 6

Cases to which (4.1) applies involve two one-dimensional spaces with span of the form

$$\left\|\begin{matrix} a & 0 & 0 \\ 0 & a & 0 \\ 0 & 0 & b \end{matrix}\right\| .$$

The other irreducible spaces are two-dimensional. One is of the form

$$\left\|\begin{matrix} a & b & 0 \\ b & -a & 0 \\ 0 & 0 & 0 \end{matrix}\right\|$$

with the basis

$$\sqrt{2}\,\mathbf{E}_1 = \left\|\begin{matrix} 0 & 1 & 0 \\ 1 & 0 & 0 \\ 0 & 0 & 0 \end{matrix}\right\| \quad , \quad \mathbf{E}_2 = \left\|\begin{matrix} 1 & 0 & 0 \\ 0 & -1 & 0 \\ 0 & 0 & 0 \end{matrix}\right\| \; ,$$

(4.5) *applies, with* $\psi = 2\pi/3$. *The other is of the form*

$$\left\|\begin{matrix} 0 & 0 & a \\ 0 & 0 & b \\ a & b & 0 \end{matrix}\right\| \; .$$

with the basis

$$\sqrt{2}\,\mathbf{E}_1 = \left\|\begin{matrix} 0 & 0 & 1 \\ 0 & 0 & 0 \\ 1 & 0 & 0 \end{matrix}\right\| \quad , \quad \sqrt{2}\,\mathbf{E}_1 = \left\|\begin{matrix} 0 & 0 & 0 \\ 0 & 0 & 1 \\ 0 & 1 & 0 \end{matrix}\right\| \; ,$$

(4.5) *applies, this time with* $\psi = \pi/3$.

No. 7

Here, there is just one space to which (4.1) applies, of the form

$$\left\|\begin{matrix} a & 0 & 0 \\ a & a & 0 \\ 0 & 0 & a \end{matrix}\right\| \; .$$

There is one of the two-dimensional kind, of the form

$$\left\|\begin{matrix} a & 0 & 0 \\ 0 & b & 0 \\ 0 & 0 & c \end{matrix}\right\| \quad ; \quad a + b + c = 0 \; .$$

with

$$\sqrt{2}\,\mathbf{E}_1 = \left\|\begin{matrix} 1 & 0 & 0 \\ 0 & -1 & 0 \\ 0 & 0 & 0 \end{matrix}\right\| \quad , \quad \sqrt{2}\,\mathbf{E}_2 = \left\|\begin{matrix} 1 & 0 & 0 \\ 0 & 1 & 0 \\ 0 & 0 & -2 \end{matrix}\right\| \; ,$$

(4.5) *applies, with* $\psi = 2\pi/3$.

Finally, there is a three-dimensional space of the form

$$\left\|\begin{matrix} 0 & a & b \\ a & 0 & c \\ b & c & 0 \end{matrix}\right\| \; .$$

With the basis

$$\sqrt{2}\,\mathbf{E}_1 = \left\|\begin{matrix} 0 & 1 & 0 \\ 1 & 0 & 0 \\ 0 & 0 & 0 \end{matrix}\right\| \,, \quad \sqrt{2}\,\mathbf{E}_2 = \left\|\begin{matrix} 0 & 0 & 0 \\ 0 & 0 & 1 \\ 0 & 1 & 0 \end{matrix}\right\| \,,$$

and

$$\sqrt{2}\,\mathbf{E}_3 = \left\|\begin{matrix} 0 & 0 & 1 \\ 0 & 0 & 0 \\ 1 & 0 & 0 \end{matrix}\right\|$$

(4.7) *applies.*

So, this describes all of the likely choices of $\ker \mathfrak{a}$, and links each one to one of six groups $\mathfrak{R}(P_0)$. The next step is to deal with the bifurcation theory for each of the six.

5. Cases with N = 1. When $N = 1$, we are concerned with the function $\mathbf{\Phi}(y, \theta)$ discussed in Section 3, y here being a scalar variable. As indicated in (3.8), we are given that it is s smooth function satisfying

$$\frac{\partial \mathbf{\Phi}}{\partial y}(0, \theta_0) = \frac{\partial^2 \mathbf{\Phi}}{\partial y^2}(0, \theta_0) = 0 \,, \tag{5.1}$$

with $(5.1)_2$ counting as one equation.

Suppose first that (4.1) applies, so that $\mathbf{\Phi}$ is not subject to any invariance conditions. Then, what is likely is that

$$\frac{\partial^2 \mathbf{\Phi}}{\partial y \partial \theta}(0, \theta_0) \neq 0 \,, \tag{5.2}$$

for the vanishing of this would give a second equation. So we have here another use of our criterion. We can then use the implicit function theorem to solve the equilibrium equation

$$\frac{\partial \mathbf{\Phi}}{\partial y} = 0 \,, \tag{5.3}$$

locally, for θ, to get

$$\theta = \theta_0 + f(y) \,, \quad f(0) = 0 \,, \quad f'(0) = 0 \,, \tag{5.4}$$

the latter condition being obtained by calculating $f'(0)$, using $(5.1)_2$. By a similar calculation, one finds that, most likely,

$$f''(0) \neq 0 \,, \tag{5.5}$$

so, near $y = 0, f$ is either positive or negative, its graph resembling a parabola. If, say, $f < 0$, one gets two equilibrium branches for $\theta < \theta_0$, coming together at $\theta = \theta_0$, to form one smooth curve in the $y - \theta$ plane. Since $\mathfrak{N}(P_0)$ is here trivial, the orbit of a point on one of these branches consists of just one point. From the discussion in Section 3, this means that the point group there is the same as it is at the reference $y = 0,\ \theta = \theta_0$ and, from the information in Section 4, this can be any of the seven. So, this gives a bifurcation involving no change of symmetry. Also, what is likely is that $\partial^2 \mathbf{\Phi}/\partial y^2$ will be positive on one of these branches, negative on the other.

When (4.3) applies, $\mathbf{\Phi}(y, \theta)$ is restricted to be an even function of y. What is then likely is the typical pitchfork bifurcation often encountered in the literature on bifurcation theory. Briefly, here

$$y = 0 \Rightarrow \frac{\partial \mathbf{\Phi}}{\partial y} = 0 \ , \tag{5.6}$$

giving one equilibrium branch, passing smoothly through $\theta = \theta_0$. Since $y = 0$ is a fixed point for this group, the point group on it is the same as that at the reference and, from the discussion in Section 4, this could be No. 2, 3, or 5. Besides the branch $y = 0$, what is likely are the branches defined either for $\theta \geq \theta_0$ or for $\theta \leq \theta_0$, describable by an equation of the form

$$y^2 = f(\theta) \ , \quad f(\theta_0) = 0 \ , \quad f'(\theta_0) \neq 0 \ , \tag{5.7}$$

where $f(\theta)$ is a smooth function. If, say, $f'(\theta_0) > 0$ and $\theta > \theta_0$, one will have an equilibrium point on one of these branches, with an orbit consisting of two points. The order of the point group at one of these will then be half of what it is at $y = 0$. According to the Landau [7] exclusion rules, these are the bifurcations associated with likely second-order phase transitions for our Bravais lattices. Here, we do not preclude the possibility that all of these branches represent unstable equilibria, and they could not describe such transitions when they are. One can deduce something about the differences in sign of $\partial^2 \mathbf{\Phi}/\partial y^2$ on the different branches, but I won't elaborate this, which will be familiar to anyone who knows a bit about such bifurcations.

Essentially, these are textbook examples, encountered in various kinds of applications. I have elaborated them a bit, partly to illustrate how our criterion for being likely is used, to get more definite predictions. Also, we will see that some of the more complicated possibilities can be reduced to one of these cases. Further, if one looks at the analysis sketched, one sees that the smoothness assumptions on $\widehat{\phi}$ are much stronger than is here necessary; to reduce smoothness assumptions to a minimum, one is likely to need to find alternatives to these and other arguments to be used.

Another point seems worth mentioning. In the case just discussed, a symmetry argument gave us the existence of one smooth branch ($y = 0$) passing through the bifurcation point. As it does, what is likely is that, on it, $\partial^2 \mathbf{\Phi}/\partial y^2$ will change sign; if it didn't, the partial derivative of this with respect to θ would vanish at bifurcation, giving a second equation. By an old result, due to Poincaré[8,§2], this implies the existence of equilibria with $y \neq 0$, in the neighborhood of the bifurcation point, and he discussed a generalization of this result to n-dimensional problems. This, combined with formal analysis, can be used to make rigorous arguments leading to (5.7), with fairly mild continuity assumptions. Very similar ideas underlie bifurcation analysis of cases yet to be considered.

6. The case N = 2, $\psi = \pi/2$. When $N = 2$, and $\mathbf{\Phi}(y_1, y_2, \theta)$ is invariant under any of the groups $G(\psi)$ described in (4.5) and (4.6), one has

$$\lambda \frac{\partial \mathbf{\Phi}}{\partial \mathbf{y}} (\mathbf{y}, \theta) = \frac{\partial \mathbf{\Phi}}{\partial \mathbf{y}} (\lambda \mathbf{y}, \theta) , \quad \lambda \in G(\psi) , \tag{6.1}$$

with $\mathbf{y} = (y_1, y_2)$. These groups have in common the property that the only vector left invariant by the group is the null vector. From (6.1), we then infer that

$$\mathbf{y} = 0 \Rightarrow \frac{\partial \mathbf{\Phi}}{\partial \mathbf{y}} = 0 , \tag{6.2}$$

so there is at least this equilibrium branch, passing smoothly through the bifurcation point.

When $\psi = \pi/2$, it is easy to verify that

$$\mathbf{\Phi}(y_1, y_2) = \mathbf{\Phi}(-y_1, y_2) = \mathbf{\Phi}(y_1, -y_2) = \mathbf{\Phi}(y_2, y_1) , \tag{6.3}$$

from which we have, in particular,

$$y_1 = 0 \Rightarrow \frac{\partial \mathbf{\Phi}}{\partial y_1} = 0 , \tag{6.4}$$

and

$$y_2 = 0 \Rightarrow \frac{\partial \mathbf{\Phi}}{\partial y_2} = 0 , \tag{6.5}$$

and (6.5) can be obtained by transforming (6.4) by an element of $G(\pi/2)$. With (6.4), and $y_1 = 0$, the equilibrium equations reduce to

$$\frac{\partial \mathbf{\Phi}}{\partial y_2} (0, y_2, \theta) = 0 , \tag{6.6}$$

$\mathbf{\Phi}$ then being an even function of y_2. Essentially, this is the second case discussed for $N = 1$. Thus we get the analog of (5.7), branches described by

$$y_1 = 0 , \quad y_2^2 = f(\theta) , \quad f(\theta_0) = 0 , \quad f'(\theta_0) \neq 0 . \tag{6.7}$$

Transforming this gives the solution of (6.5)

$$y_2 = 0, \quad y_1^2 = f(\theta) . \tag{6.8}$$

Clearly, where f is positive, the orbit of one equilibrium point like $\left(0, \sqrt{f(\theta)}\right)$, consists of four distinct points. From Section 4, the point group at $\mathbf{y} = 0$, $\theta = \theta_0$ can only be No. 4, configurations of the tetragonal kind, so (6.2) is a branch with this symmetry. From the Appendix, the order of the point group is eight. That on the symmetry-related branches described by (6.7) and (6.8) is then $8/4 = 2$, making this the point group No. 2, the monoclinic configurations.

So, we have at least these branches. To explore the possibility of additional branching, we use (6.3), along with a little exercise in invariance theory, to infer that

(6.9) $$\mathbf{\Phi} = F(I, J, \theta)$$

where F is a smooth function of the variables

(6.10) $$I = |\mathbf{y}|^2 = y_1^2 + y_2^2 \ , \qquad J = y_1^2 \, y_2^2 \ .$$

While our smoothness assumptions make it relevant, it is worth noting that, were $\mathbf{\Phi}$ only a few times differentiable, F would be a bit less smooth, at some places, as follows from the results of Ball [9]. For any equilibrium points satisfying the condition

(6.11) $$\frac{\partial(I, J)}{\partial(y_1, y_2)} = 4 y_1 y_2 (y_1^2 - y_2^2) \neq 0 \ ,$$

the equilibrium equations become

(6.12) $$\frac{\partial F}{\partial I} = \frac{\partial F}{\partial J} = 0 \ .$$

Suppose that there were a sequence of such equilibrium points, having $\mathbf{y} = 0$, $\theta = \theta_0$ as a limit point. Then, by continuity, we would have

(6.13) $$\frac{\partial F}{\partial I} \, (0, 0, \theta_0) = \frac{\partial F}{\partial J} \, (0, 0, \theta_0) = 0 \ ,$$

that is, at $\mathbf{y} = 0$, $\theta = \theta_0$. To see whether this is likely, we can use the first few terms in an expansion of $\mathbf{\Phi}$ about $\mathbf{y} = 0$, *viz.*

(6.14) $$\mathbf{\Phi} = \alpha + \beta I + \gamma J + \delta I^2 + o(|\mathbf{y}|^4) \ ,$$

where α, β *etc.* are functions of θ. From our assumptions one equation must be satisfied, *viz.*

(6.15) $$\beta(\theta_0) = 0 \ ,$$

so it is unlikely that $\gamma(\theta_0) = 0$, *i.e.* that (6.13) is satisfied. So, we expect not to have equilibria of this kind. From (6.11), and our coverage of cases where $y_1 y_2 = 0$, this leave the possibility of equilibria such that $y_1^2 = y_2^2 \neq 0$ and, by similar arguments, one can show that these are unlikely to occur.

7. The case $N = 2$, $\psi = 2\pi/3$. With $\mathbf{\Phi}(y_1, y_2, \theta)$ invariant under the group $G(2\pi/3)$ described by (4.5) and (4.6), we already have from (6.2), the branch

$$\mathbf{y} = 0 \ , \tag{7.1}$$

and the easy observation that

$$y_1 = 0 \Rightarrow \frac{\partial \mathbf{\Phi}}{\partial y_1} = 0 \ . \tag{7.2}$$

This suggests looking at the possibility of satisfying

$$\frac{\partial \mathbf{\Phi}}{\partial y_2} \ (0, y_2, \theta) = 0 \ , \tag{7.3}$$

with $y_2 \neq 0$. While this might seem to be the same as one of the two cases for which $N = 1$, it is not. For one thing, one can differentiate (6.1) with respect to θ to get, in particular,

$$\frac{\partial^2 \mathbf{\Phi}}{\partial y_2 \partial \theta} \ (0, 0, \theta) = 0 \ , \tag{7.4}$$

a consequence of symmetry. This makes it clear that this is unlike the first case discussed in Section 5. Also, it is easy to check that $\mathbf{\Phi}(0, y_2, \theta)$ is here not an even function of y_2, making it unlike the second case.

By using our now-familiar criterion, it is easy to check that, most likely,

$$\frac{\partial^3 \mathbf{\Phi}}{\partial y_2^2 \partial \theta} \ (0, 0, \theta_0) \neq 0 \ . \tag{7.5}$$

So, on the branch (7.1), $\partial^2 \mathbf{\Phi} / \partial y_2^2$ will change sign, as θ passes through the value θ_0. By applying the result of Poincaré, mentioned at the end of Section 5, to $\mathbf{\Phi}(0, y_2, \theta)$, one infers the existence of solutions of (7.3) with $y_2 \neq 0$, in the neighborhood of $y_2 = 0$, $\theta = \theta_0$. I won't expend the work required to describe my elementary analysis of these, but it produces one branch described by

$$y_1 = 0 \ , \quad y_2 = f(\theta) \ , \quad f(\theta_0) = 0 \ , \quad f'(\theta_0) \neq 0 \ . \tag{7.6}$$

Transforming this by elements of the group gives two more, of the form

$$2y_1 = \pm\sqrt{3} \ f(\theta) \ , \quad 2y_2 = -f(\theta) \ , \tag{7.7}$$

or three symmetry-related branches in all.

To explore whether occurrence of additional branches is likely, we note that the invariance of $\mathbf{\Phi}$ under $G(2\pi/3)$ implies that it is an even function of y_1, with y_2 fixed. Thus, we can regard it as smooth function of y_1^2 and y_2 or, equivalently, of the variables

$$X = y_1^2 - 3y_2^2 \ , \ Y = y_2 \ , \tag{7.8}$$

so

$$\Phi(y_1, y_2, \theta) = F(X, Y, \theta) \ , \tag{7.9}$$

a function which is still subject to some invariance requirements. What we have above are equilibria on the lines $y_1 = 0$, and $y_1 = \pm\sqrt{3}\ y_2$, so let us look for equilibria occurring elsewhere with, say, $\theta > \theta_0$. If $f(\theta)$ is then positive, we look at the domain described by

$$y_2 < 0 \ , \quad \sqrt{3}\ y_2 < y_1 < 0 \ . \tag{7.10}$$

If we have any equilibrium point not on one of the three lines, some point on its orbit will be in this domain. In the $X-Y$ plane, this maps to

$$Y < 0 \ , \quad -3Y^2 < X < 0 \ . \tag{7.11}$$

Here, we are interested in possible solutions of

$$\frac{\partial F}{\partial X} = \frac{\partial F}{\partial Y} = 0 \tag{7.12}$$

and know that they are satisfied at $(0, -f(\theta)/2)$. Suppose that, at $\theta = \theta_1$, they are also satisfied at $(X_1, Y_1) \neq (0, -f(\theta_1)/2)$ and consider the function

$$g(\tau) = \frac{\partial F}{\partial X}\ \left[\tau X_1 \ , \ \tau Y_1 + (\tau - 1) f(\theta_1)/2, \theta_1\right] \ . \tag{7.13}$$

It vanishes at $\tau = 0$ and $\tau = 1$ so, for some $\tau \in (0,1), g'(\tau) = 0$. This gives us a point at which

$$\frac{\partial^2 F}{\partial X^2}\ X_1 + \frac{\partial^2 F}{\partial X \partial Y}\ (Y_1 + f(\theta_1)/2) = 0 \ . \tag{7.14}$$

In a similar way, by using $\partial F/\partial Y$, we can get a point, generally different, at which

$$\frac{\partial^2 F}{\partial X \partial Y}\ X_1 + \frac{\partial^2 F}{\partial Y^2}\ (Y_1 + f(\theta_1)/2) = 0 \ . \tag{7.15}$$

Here, we can replace $(X_1, Y_1 + F(\theta_1)/2)$ by the unit vector

$$(X_1/r,\ (Y_1 + f(\theta_1)/2)/r) \ , \ r = |(X_1, Y_1 + f(\theta_1)/2)| \ . \tag{7.16}$$

Now, suppose that there were a sequence of such equilibria, having as a limit point (0,0) as $\theta \to \theta_0$. This will give a sequence of points on the unit circle which must have some limit point. Also, the points where (7.14) and (7.15) hold will tend to (0,0) as $\theta \to \theta_0$. Putting this together, one gets the condition

$$\frac{\partial^2 F}{\partial X^2}\ \frac{\partial^2 F}{\partial Y^2} - \left(\frac{\partial^2 F}{\partial X \partial Y}\right)^2 = 0 \quad \text{at} \quad X = Y = 0 \ , \quad \theta = \theta_0 \ . \tag{7.17}$$

Now, with a little exercise in invariant theory, one gets the approximation

$$\mathbf{\Phi} = \alpha + \beta|\mathbf{y}|^2 + \gamma y_2\,(3y_1^2 - y_2^2) + \delta|\mathbf{y}|^4 + o(|\mathbf{y}|^4)$$

or

$$F(X,Y,\theta) \cong \alpha + \beta(X+4Y^2) + \gamma Y(3X-4Y^2) + \delta(X+4Y^2)^2\ , \tag{7.18}$$

where α, β, *etc.* are functions of θ. From our assumptions, $\beta(\theta_0)$ must vanish, one equation to be satisfied. For (7.17) to hold would require that $\gamma(\theta_0)$ also vanish, so this is unlikely. With slight changes in the argument, one can cover the possibility that $f(\theta) < 0$ for $\theta > \theta_0$, and cover equilibria which might occur for $\theta < \theta_0$. In brief, what is likely is that we will have the branches described by (7.1), (7.6) and (7.7), and no others. Physically, these could be the trigonal-monoclinic, hexagonal-orthorhombic or cubic-tetragonal kind. The later possibility is discussed in more detail in Section 10.

8. The case $N = 2$, $\psi = \pi/3$. With $\mathbf{\Phi}(y_1, y_2, \theta)$ invariant under $G(\pi/3)$, we again have the trivial branch

$$\mathbf{y} = 0\ , \tag{8.1}$$

for a start. Notice that this group includes rotations with angle π, and you see that $\mathbf{\Phi}$ satisfies

$$\mathbf{\Phi}(y_1, y_2, \theta) = \mathbf{\Phi}(-y_1, y_2, \theta) = \mathbf{\Phi}(y_1, -y_2, \theta)\ , \tag{8.2}$$

as well as some additional conditions. One oddity is that one has to go to approximations of rather high order, to get polynomials which are invariant under this group, and not the full orthogonal group; my calculations give

$$\mathbf{\Phi} = \alpha + \beta|\mathbf{y}|^2 + \gamma|\mathbf{y}|^4 + \delta|\mathbf{y}|^6 + \varepsilon(y_1^2 - y_2^2)\,[4(y_1^2 - y_2^2) - 3|\mathbf{y}|^4] + o(|\mathbf{y}|^6)\ , \tag{8.3}$$

where α, β, *etc.* are functions of θ. Still, $\mathbf{\Phi}(0, y_2, \theta)$ is a rather general even function of y_2 and, as was the case for $G(\pi/2)$, we get branches of the typical pitchfork kind

$$y_1 = 0\ , \quad y_2^2 = f(\theta)\ , \quad f(\theta_0) = 0\ , \quad f'(\theta_0) \neq 0\ . \tag{8.4}$$

Transforming this by the elements of the group has the effect of rotating this by angles, $n\pi/3$, $n = 1,2,3,4,5$. Since (8.3) describes two equilibria when $f > 0$, the orbit consists of twelve distinct points. Checking the information you have been given, in Section 4 and the Appendix, you find that, so far, what we have is a hexagonal-triclinic bifurcation. By essentially the same reasoning, one can get a similar but different array of branches, in the orbit of one of the form

$$y_2 = 0\ , \quad y_1^2 = g(\theta)\ , \quad g(\theta_0) = 0\ , \quad g'(\theta_0) \neq 0\ . \tag{8.5}$$

a fact which seems to me rather curious.

Here, it is convenient to regard $\mathbf{\Phi}$ as a somewhat special function of the variables

(8.6) $$K = |\mathbf{y}|^2 \ , \quad L = y_1^2 - y_2^2 \Rightarrow |L| \le K \ , \quad K > 0 \ .$$

and θ,

(8.7) $$\mathbf{\Phi} = F(K, L, \theta) \ ,$$

(8.2) serving to justify this. In terms of these, some of the equilibria described above lie on the lines in the $K - L$ plane given by

(8.8) $$L = \pm K \ .$$

If there are any other kinds, they will be in the sector

(8.9) $$K > 0 \ , \quad |L| < K \ .$$

Suppose that there is one there, at (K_0, L_0), at $\theta = \theta_1$. By a routine calculation to determine others on the same orbit, one gets them at

(8.10) $$(K_0, -L_0/2 \pm M_0) \ , \ 4M_0^2 = 3\ (K_0^2 - L_0^2) \ .$$

Now, when the three points indicated are distinct, repeated use of the mean value theorem yields the condition

$$\frac{\partial^3 F}{\partial L^3}\ (K_1, L_1, \theta_1) = 0 \ ,$$

(K_1, L_1) being some point in the sector. So, if there is a sequence of these approaching $(0,0)$ as $\theta_1 \to \theta_0$, we must have

(8.12) $$\frac{\partial^3 F}{\partial L^3}\ (0, 0, \theta_0) = 0 \ .$$

With (8.3), this gives $\varepsilon(\theta_0) = 0$ and, by now familiar reasoning, this is unlikely.

Within the sector, $M_0 \neq 0$, so, for two of the points to coincide, we must have

(8.13) $$L_0 = -L_0/2 \pm M_0 \ ,$$

in which case there are two distinct points. By a routine calculation, these correspond to points on the lines

$$y_1 = \pm\sqrt{3}\ y_2 \ , \quad y_2 = \pm\sqrt{3}\ y_1$$

lines obtained by taking orbits of the lines $y_1 = 0$ and $y_2 = 0$. So, there are such equilibria, but they have been accounted for.

Here, particularly, I find the analysis unpleasant, in that they presume much differentiability of $\widehat{\phi}$. Perhaps some nicer treatment is in the literature, but I haven't found one.

9. The case N = 3. Here, we are dealing with functions $\mathbf{\Phi}(y_1, y_2, y_3, \theta)$, which are invariant under the large group $G(3)$ with generators described in (4.8). Again, the only vector left invariant by this group is the null vector, so we have the branch implied by

$$\mathbf{y} = 0 \Rightarrow \frac{\partial \mathbf{\Phi}}{\partial \mathbf{y}} = 0 \, , \tag{9.1}$$

passing smoothly through $\mathbf{y} = 0$, $\theta = \theta_0$, configurations of the cubic kind. The invariance implies that, when any one of the coordinates vanishes, $\mathbf{\Phi}$ reduces to an even function of the other two, whence follows that

$$y_2 = y_3 = 0 \Rightarrow \frac{\partial \mathbf{\Phi}}{\partial y_2} = \frac{\partial \mathbf{\Phi}}{\partial y_3} = 0, \; etc. \tag{9.2}$$

with $\mathbf{\Phi}(y_1, 0, 0, \theta)$ an even function of y_1. After checking the invariance requirements on it, one finds that these don't exclude the occurrence of the typical pitchfork bifurcations, so we will have branches described by

$$y_2 = y_3 = 0 \, , \quad y_1^2 = f(\theta) \, , \quad f(\theta_0) = 0 \, , \quad f'(\theta_0) \neq 0 \, . \tag{9.3}$$

The orbit of this gives two more "copies," obtained by replacing y_1 by y_2 and by y_3. So, where $f(\theta) > 0$, the orbit of one such equilibrium point contains six distinct points. The order of the cubic group is 24, so the order of the point on any of these branches is 24/6=4, identifying this as the orthorhombic group. Here, at least when $\mathbf{\Phi}$ is a polynomial, it is known that it can be shown to be representable in the form

$$\mathbf{\Phi} = \mathbf{Q}F_1 + \mathbf{K}\mathbf{\Phi}_2 \, , \quad \mathbf{K} = y_1 y_2 y_3 \, , \tag{9.4}$$

with $\mathbf{\Phi}_1$ and $\mathbf{\Phi}_2$ symmetric functions of y_1^2, y_2^2 and y_3^2, expressible as functions of

$$\left. \begin{aligned} I &= y_1^2 + y_2^2 + y_3^2 \, , \\ J &= y_1^4 + y_2^4 + y_3^4 \, , \\ \text{and} \quad K^2 &= (y_1 y_2 y_3)^2 \, . \end{aligned} \right\} \tag{9.5}$$

Or, more simply, we have

$$\mathbf{\Phi} = F(I, J, K, \theta) \tag{9.6}$$

where F is a polynomial in I, J, K*. In fact, one can adapt the analysis by Ball [9] to justify our assuming that (9.6) holds, with F a smooth function. At points such that

$$\frac{\partial(I, J, K)}{\partial(y_1, y_2, y_3)} = 8(y_1^2 - y_2^2)\,(y_2^2 - y_3^2)\,(y_3^2 - y_1^2) \neq 0 \tag{9.7}$$

*This can be inferred from results of Smith and Rivlin [10], for example.

the equilibrium equations are

$$\frac{\partial F}{\partial I} = \frac{\partial F}{\partial J} = \frac{\partial F}{\partial K} = 0 \,, \tag{9.8}$$

and it is easy enough to show that it is unlikely that these hold at $\mathbf{y} = 0$, $\theta = \theta_0$, so, by reasoning now familiar, we can dismiss such equilibria. From (9.7), this leaves unsettled the possibilities like

$$y_2 = \pm y_3 \neq 0 \,. \tag{9.9}$$

From above, it is easy to see that the restriction of $\mathbf{\Phi}$ to the set

$$y_2 = y_3 \,, \tag{9.10}$$

say, is an even function of this variable, so we can regard it as a function of its square, so

$$\mathbf{\Phi}(y_1, y_2, y_2, \theta) = G(X, Z, \theta) \tag{9.11}$$

with

$$X = y_1 \,, \quad Z = y_2^2 \,. \tag{9.12}$$

From, say, (9.5) and (9.6), it is clear that

$$\mathbf{\Phi} = \alpha + \beta I + \gamma K + \delta I^2 + \varepsilon J + o(|y|^4) \,, \tag{9.13}$$

where $\alpha\beta$, *etc.*are functions of θ,

$$\beta(\theta_0) = 0 \tag{9.14}$$

being the one equation we allow. From this we have the approximation

$$G \cong \alpha + \beta(X^2 + 2Z) + \gamma X K + \delta(X^2 + 2Z)^2 + \varepsilon(X^4 + 2Z^2) \,. \tag{9.15}$$

Also, for equilibria with $Z > 0$, conditions of equilibria imply that

$$\frac{\partial G}{\partial X} = \frac{\partial G}{\partial Z} = 0 \,. \tag{9.16}$$

With (9.15), it is easy to verify that these are satisfied at $X = 0$, $Z = 0$, $\theta = \theta_0$. Also, here,

$$\det \left\| \begin{matrix} \dfrac{\partial^2 G}{\partial X^2} & \dfrac{\partial^2 G}{\partial X \partial Z} \\ \\ \dfrac{\partial^2 G}{\partial X \partial Z} & \dfrac{\partial^2 G}{\partial Z^2} \end{matrix} \right\| = -\gamma^2(\theta_0) \,, \tag{9.17}$$

which is not likely to vanish. So, the implicit function theorem applies, giving us the existence and uniqueness of solutions of (9.16), locally. In first approximation, we get the solution as

$$\left.\begin{aligned} X &\cong -2\beta'(\theta_0)(\theta-\theta_0) \quad / \quad \gamma(\theta_0), \\ Z &\cong 4[\beta'(\theta_0)]^2(\theta-\theta_0)^2 \quad / \quad \gamma(\theta_0)^2, \end{aligned}\right\} \tag{9.18}$$

which is consistent with having $Z > 0$, as it should be. Really, there are three equilibrium equations to be satisfied, in general. However, it is easy to check that

$$y_1 = y_1 \Rightarrow \frac{\partial \Phi}{\partial y_1} = \frac{\partial \Phi}{\partial y_2} \; ; \tag{9.19}$$

one can use (9.6) to confirm this, for example, as well as checking that it is sufficient to satisfy (9.16). Now (9.18) suggests that, for these solutions, $Z = X^2$, and it isn't hard to verify this, using the uniqueness of solutions. So, this gives us equilibria of the form

$$y_1 = f(\theta) \;, \quad y_2 = y_3 \quad y_2^2 = f^2(\theta) \;, \tag{9.20}$$

(9.18) giving a first approximation of $f(\theta)$. Of course, it is unlikely that $\beta'(\theta_0) = 0$. Naturally, $G(3)$ generates an orbit of "copies" of (9.20), which covers the possibilities of having any two of the three quantities y_1^2, y_2^2, y_3^2 be equal and nonzero, and we have also covered the cases where some or all vanish. Counting up the number of distinct points on an orbit of this kind, I find four. So the order of the point group for these is 24/4=6, the trigonal type. So, we have quite a variety of branches, in this case. I do not know of observations of a softening modulus which would suggest a bifurcation of this kind. However, there aren't very many observations of this general kind, and there are hardly any for Martensitic phases, which here could be either of the orthorhombic or the trigonal kind, for suitable forms of the function $\widehat{\phi}$.

10. Example. For shape-memory alloys, one of the more commonly observed possibilities has as Austenite a crystal of cubic form. Since this group is maximal, any limit point is also cubic. Upon cooling it, the observations indicate softening described by

$$C_{11} - C_{12} \to 0 \tag{10.1}$$

in notation commonly used by experimentalists and by Love [11], for example. Here the C's are not to be confused with components of the tensor $\mathbf{C}$. Referring to the comments about the cubic (No. 7) in Section 4, the eigenvalue which seems to vanish is that associated with the two-dimensional eigenspace, giving us the case $N = 2$, $\mathfrak{R}(P_0) = G(2\pi/3)$ considered in Section 7: the observations give that the other two eigenvalues remain positive. From the discussion in Section 7, the cubic branch passes smoothly through $\theta = \theta_0$, with the indicated eigenvalue becoming negative, for $\theta < \theta_0$, or, said differently,

$$C_{11} - C_{12} < 0 \quad \text{for} \quad \theta < \theta_0 \;. \tag{10.2}$$

Also, we have the three symmetry-related branches noted in (7.6) and (7.7), involving the tetragonal symmetry. One of these will match the description in the Appendix, with the same basis $\mathbf{e}_1, \mathbf{e}_2, \mathbf{e}_3$ use for the tetragonal and the cubic configurations. From Section 4, $\mathfrak{a}$ generally has five distinct values for tetragonal crystals, but at the bifurcation point one has the cubic symmetry so, as a consequence of invariance, some of the five must become equal to match the three for the cubic. This provides an example of an exceptional case, not regarded as unlikely, in which normally distinct eigenvalues can become equal and also vanish. Looking at what is involved, we find that, for the tetragonal, the eigenvalues associated with the two-dimensional eigenspace and the one-dimensional space of the form

$$\left\| \begin{matrix} 0 & a & 0 \\ a & 0 & 0 \\ 0 & 0 & 0 \end{matrix} \right\| \tag{10.3}$$

become equal to a positive eigenvalue for the cubic, so both of these are positive, locally. Of the two eigenvalues with span of the form

$$\left\| \begin{matrix} a & 0 & 0 \\ 0 & a & 0 \\ 0 & 0 & b \end{matrix} \right\| \tag{10.4}$$

one approaches that associated with the one-dimensional eigenspace for the cubic and is therefore positive. The other merges with that for which the eigenspace is of the form

$$\left\| \begin{matrix} a & 0 & 0 \\ 0 & -a & 0 \\ 0 & 0 & 0 \end{matrix} \right\| \tag{10.5}$$

these vanishing at the bifurcation point. With some additional analysis, which I won't elaborate, one can show that, nearby, one of the two is positive, one negative, with both of these changing sign as θ passes through θ_0, indicating that these branches are unstable, having one negative eigenvalue, associated with a one-dimensional eigenspace. Actually, that associated with (10.5) is positive for $\theta > \theta_0$. There are simple possibilities for such branches regaining stability. For example, for $\theta > \theta_0$, the negative eigenvalue is one of those with the span (10.3), and it could become zero at some such temperature. From Section 4, this would give a bifurcation for which $N = 1$, the first kind discussed in Section 5. That is, one would have a tetragonal branch with positive eigenvalues, for temperatures below the temperature at which this bifurcation occurs. Exchange of stability between this and the cubic would then produce a rather typical subcritical bifurcation, a cubic-tetragonal phase transition of first order, and transitions of this general kind occur in shape-memory alloys. If this picture applies, one has a prediction that a certain eigenvalue associated with the tetragonal should become relatively small, near the indicated bifurcation. It is at least within the realm of possibility that experiments could confirm or deny this: I have not seen such data yet, but expect that experimentalists will produce some eventually. It isn't hard to see that more complicated combinations of bifurcations can be associated with cubic-tetragonal transitions and predict softening of the same tetragonal modulus. Simply, it is not very clear what can be done, experimentally, to discern between such possibilities, but this is a question which seems to deserve thoughtful consideration.

With the information provided, an interested reader can similarly treat the possibility that the negative eigenvalue which again vanishes is that associated with (10.5), which gives a theoretically possible picture of a cubic-orthorhombic transition of first order, associated with softening of the same cubic modulus. Again, there are observations of cubic-orthorhombic transitions associated with this kind of softening in shape-memory alloys, but there is a quirk. What the suggested analysis gives could fit some kinds of cubic-orthorhombic transitions, but not one of the body-centered-cubic-face-centered-orthorhombic type, for example; this can be inferred by calculating the form of $\mathbf{C}$ associated with the orthotropic phase. On the other hand, transitions of this kind are observed. If nothing else, this indicates that it can be feasible to prove wrong a simple guess about a global bifurcation pattern, with the rather limited experimental information now available. More accurately, I am inclined to believe that the general theory does apply to such transitions, it being the guess about the global bifurcation pattern which is at fault. After seeing what is wrong, one can make better educated guesses about the global pattern, finding some which could fit such transitions. One might find reason to reject some such, given experimental information concerning what is the soft modulus in the orthorhombic phase.

This discussion serves to illustrate some of the kinds of things that can be inferred about unstable branches, using kinds of experimental results which could now be obtained, by knowledgeable experimentalists.

I see this as a small piece of a larger problem. Among workers interested in phase transitions, it is a rather common practice to make some guess about the form of the relevant thermodynamic potential, involving some adjustable parameters which can be determined, using data which either has been or obviously can be obtained. Roughly, the general problem is to find the qualities characterizing the more likely guesses, and I expect some of these to be of a qualitative nature. For the more mathematical studies, what is often important is not the specific form of our function $\widehat{\phi}$, say, but of its qualitative features. Looking back at our analyses, we have used some polynomial approximations of fairly high degree, and such polynomials represent popular guesses as to the form of $\widehat{\phi}$. From the work of Smith and Rivlin [10], one can determine those which are invariant under any one of the point groups. My experience is that you will have trouble finding relevant data to determine all of the adjustable constants involved. Even if you find enough data to determine all, you are likely to have a complicated function, making it hard to see just what general properties it has, that might enable you to apply some mathematical theorem. Thinking about this persuaded me that it was worthwhile to develop the kinds of pictures which I have presented here. Those who wonder why I have not considered crystals of more than general kinds might consult the paper by Zanzotto [12]; there are some good physical reasons for this.

Appendix

The (three-dimensional) crystallographic groups, being subgroups of the orthogonal group $O(3)$, have elements $\mathbf{Q} \in O(3)$, representable in the form

$$\mathbf{Q} = \pm\mathbf{R} ,$$

where

$$\mathbf{R} = \mathbf{R}(\psi, \mathbf{e}) \in SO(3) ,$$

is a rotation with angle $\psi, \mathbf{e}$ being a unit vector, representing the axis of rotation. The point groups realizable in Bravais lattices all contain the central inversion $\mathbf{Q} = -\mathbf{1}$, so can be generated by this and some group consisting of rotations. What is referred to as the point group P_0 in this paper is the latter group. Strictly speaking, this is not correct, but I don't think that this will cause confusion.

Involved in the description of each is an orthonormal basis $\mathbf{e}_i$ $(i = 1, 2, 3)$, vectors often referred to as crystallographic axes. Where you see the designation $\mathbf{e} = \mathbf{p}$, this means to use for $\mathbf{e}$ all of the four vectors

$$(\mathbf{e}_1 \pm \mathbf{e}_2 \pm \mathbf{e}_3)/\sqrt{3} .$$

Listed according to their order, the indicated groups are

Number	Elements	Order
1	$\mathbf{1}$	1
2	$\mathbf{1}, \mathbf{R}(\pi, \mathbf{e}_3)$	2
3	$\mathbf{1}, \mathbf{R}(\pi, \mathbf{e}_1), \mathbf{R}(\pi, \mathbf{e}_2), \mathbf{R}(\pi, \mathbf{e}_3)$	4
4	$\mathbf{1}, \mathbf{R}(\pi, \mathbf{e}_1), \mathbf{R}\left(\pi, \dfrac{\mathbf{e}_1 \pm \sqrt{3}\,\mathbf{e}_2}{2}\right)$, $\mathbf{R}(2\pi/3, \mathbf{e}_3), \mathbf{R}(4\pi/3, \mathbf{e}_3)$	6

Number	Elements	Order
5	$\mathbf{1},\ \mathbf{R}(\pi,\mathbf{e}_1),\mathbf{R}(\pi,\mathbf{e}_2),\mathbf{R}(\pi,\mathbf{e}_3),$ $\mathbf{R}\left(\pi,\dfrac{\mathbf{e}_1\pm\mathbf{e}_2}{\sqrt{2}}\right),\mathbf{R}(\pi/2,\ \mathbf{e}_3),\mathbf{R}(3\pi/2,\mathbf{e}_3)$	8
6	$\mathbf{1},\ \mathbf{R}(\pi,\mathbf{e}_1),\mathbf{R}(\pi,\mathbf{e}_2),\mathbf{R}(\pi,\mathbf{e}_3),$ $\mathbf{R}\left(\pi,\dfrac{\mathbf{e}_1\pm\sqrt{3}\,\mathbf{e}_2}{2}\right),\mathbf{R}\left(\pi,\dfrac{\sqrt{3}\,\mathbf{e}_1\pm\mathbf{e}_2}{2}\right),\mathbf{R}(\pi/3,\mathbf{e}_3),$ $\mathbf{R}(2\pi/3,\mathbf{e}_3),\mathbf{R}(4\pi/3,\mathbf{e}_3),\mathbf{R}(5\pi/3,\mathbf{e}_3)$	12
7	$\mathbf{1},\ \mathbf{R}(\pi/2,\mathbf{e}_1),\mathbf{R}(\pi/2,\mathbf{e}_3),\mathbf{R}(\pi/2,\mathbf{e}_3)$ $\mathbf{R}\left(\pi,\dfrac{\mathbf{e}_1\pm\mathbf{e}_2}{\sqrt{2}}\right),\mathbf{R}\left(\pi,\dfrac{\mathbf{e}_2\pm\mathbf{e}_3}{\sqrt{2}}\right),\mathbf{R}\left(\pi,\dfrac{\mathbf{e}_1\pm\mathbf{e}_3}{\sqrt{2}}\right),$ $\mathbf{R}(\pi,\mathbf{e}_1),\mathbf{R}(\pi,\mathbf{e}_2),\mathbf{R}(\pi,\mathbf{e}_3),$ $\mathbf{R}(3\pi/2,\mathbf{e}_1),\mathbf{R}(3\pi/2,\mathbf{e}_2),\mathbf{R}(\pi/2,\mathbf{e}_3),$ $\mathbf{R}(2\pi/3,\mathbf{p}),\mathbf{R}(4\pi/3,\mathbf{p}),$	24

Crystals having symmetries described by one of these forms one of the crystal systems or classes, as they are sometimes called. Different workers use different names for these, those used here being as follows:

1. Triclinic
2. Monoclinic
3. Orthorhombic
4. Trigonal
5. tetragonal
6. Hexagonal
7. Cubic

One can further refine these into different symmetry types. As is discussed by Ericksen [4], this is of some important for understanding what kinds of bifurcations are possible, but here, it is not necessary to make any explicit use of these refinements. One could specify the symmetry type at bifurcation, then infer something about the symmetry types on branches, for any of the cases here considered. By not doing so, I have left the theory somewhat incomplete. For those familiar with such matters, I don't think that it would be hard to fill in details of this kind.

Acknowledgement. This work was supported by AFOSR/XOP and NSF, under grant NSF/DMS 8718881.

REFERENCES

[1] *Material Instabilities in Continuum Mechanics* (ed. J.M. Ball), Oxford University Press (1988).

[2] *Shape Memory Effects in Alloys* ed. J. Perkins), Plenum Press, New York–London (1975).

[3] Ericksen, J.L., *Bifurcation and Martensitic transformations in Bravais lattices*, to appear in J. Elasticity.

[4] Ericksen, J.L., *Weak Martensitic transformations in Bravais lattices*, Arch. Ration. Mech. Anal. **107** (1989), pp. 23–36.

[5] Truesdell, C., *First Course in Rational Mechanics*, vol. 1, Academic Press, New York–San Francisco–London (1977).

[6] Man, C.-S., *Material stability, the Gibbs conjecture and the first phase rule for substances*, Arch. Ration. Mech. Anal. **91** (1985), pp. 1–53.

[7] Landau, L.D., *On the theory of phase transitions*, in *Collected Papers of L.D. Landau* (ed. D. TerHaar) Gordon and Breach, and Pergamon Press, New York–London–Paris (1965).

[8] Poincaré, H., *Sur l'Équilibre d' une mass fluide animée d' une mouvement de rotation*, Acta Math. **7** (1885), pp. 259–380.

[9] Ball, J.M., *Differentiability properties of symmetric and isotropic functions*, Duke Math. J. **51** (1984), pp. 699–728.

[10] Smith, G.F. and Rivlin, R.S., *The strain-energy function for anisotropic elastic materials*, Trans. Am. Math. Soc. **88** (1958), pp. 175–193.

[11] Love, A.E.H., *A Treatise on the Mathematical Theory of Elasticity*, 4th ed., Cambridge University Press (1927).

[12] Zanzotto, G., *Twinning in minerals and metals: remarks on the comparison of a thermoelasticity theory with some available experimental results*, Notes I and II, Atti Acc. Lincei Rend. Fis. **82** (1988), pp. 723–741 and 743–746.

GEOMETRIC PARAMETERS AND THE RELAXATION OF MULTIWELL ENERGIES

NIKAN B. FIROOZYE*† AND ROBERT V. KOHN**‡

Abstract. This paper discusses the relaxation of a multiwell energy of the special form $W = \min_i\{|\nabla u - a^i|^2\}$. We explain how the relaxation QW can be expressed in terms of certain "tensors of geometric parameters." The exact set $\mathcal{F}_\theta$ of attainable geometric parameters is not known, but we show that it must lie inside an explicitly given convex set $\mathcal{F}_\theta^U$. This leads to a new Geometric Parameters Lower Bound for QW. For the special case of three wells in two space dimensions we give a complete characterization of the extreme points of $\mathcal{F}_\theta^U$. The final section addresses the "three gradient problem," which asks whether three pairwise incompatible gradients can nevertheless be mutually compatible. We do not solve this problem, but we show that it is linked to the attainability of the type 3 extreme points of $\mathcal{F}_\theta^U$.

1. Introduction. A basic problem in the variational modeling of coherent phase transitions is the identification of energy-minimizing microstructures, see e.g. [4,5,14,15,17,25]. Mathematically speaking, this is equivalent to the relaxation of a multiwell energy $W(\nabla u)$ of the form

$$W(\nabla u) = \min_{i=1,\ldots,N}\{W_i(\nabla u)\}, \tag{1.1}$$

see e.g. [17]. The individual "wells" $\{W_i(\nabla u)\}$ are the energies of the N component phases. The relaxed or "macroscopic" energy is QW, the quasiconvexification of W, defined by

$$QW(\xi) = \inf_{u|_{\partial\Omega}=\xi\cdot x} \frac{1}{|\Omega|}\int_\Omega W(\nabla u)dx. \tag{1.2}$$

It represents the average energy of the system when the average gradient is ξ, assuming that both the phase geometry and the locally varying deformation gradient are governed by energy minimization. The process of relaxation is discussed, for example, in [1,7,9,18,19].

To calculate the relaxed energy, one must find a formula that is at once an upper bound and a lower bound for $QW(\xi)$. Upper bounds are obtained by considering specific microstructures, obtained for example by "sequential lamination." (The best such upper bound is the "rank-one convexification," see [8,18,23].) Lower bounds are usually established by some version of the "translation method," which is based on the use of weakly lower semicontinuous functions and convexification

---

*Institute for Mathematics and its Applications.

†The work of NBF was supported in part by the Institute for Mathematics and its Applications with funds provided by the National Science Foundation and the U.S. Army Research Office. An earlier version of this work appeared as part of NBF's Ph.D. Thesis, *Optimal Translations and Relaxations of Some Multiwell Energies*, New York University, August 1990.

**Courant Institute of Mathematical Sciences.

‡The work of RVK was supported in part by NSF grant DMS-8701895, ONR grant N00014-88-K-0279, AFOSR grant 90-0090, ARO contract DAAL03-89-K-0039, and DARPA contract F49620-87-C-0065.

[12]. (The best such bound using only weakly continuous translations is the "polyconvexification," see [8,18,23].) These methods have the advantage of being quite general. However, it can be difficult to apply them optimally in specific cases. Moreover, we do not know whether the optimal translation bound agrees in general with the rank-one convexification. Therefore it is natural to seek new approaches to computing relaxed energies.

This paper explores a new method for bounding QW from below, when W has the specific form

$$W(\nabla u) = \min_{i=1,\ldots,N} \{|\nabla u - a^i|^2\} \tag{1.3}$$

for some matrices a^i, $1 \leq i \leq N$. The basic idea is as follows. To any microstructure we shall associate a "tensor of geometric parameters" $F = (F_{ij\alpha\beta})$. The effective energy of the microstructure is explicitly representable as a linear function of F. If we knew the exact set $\mathcal{F}_\theta$ of geometric parameters attained by microstructures with volume fraction θ, then we could compute $QW(\xi)$ exactly by minimizing this linear function over $\mathcal{F}_\theta$. We do not know $\mathcal{F}_\theta$ explicitly, but we do know a set $\mathcal{F}_\theta^U$ which contains it; minimization over $\mathcal{F}_\theta^U$ gives a lower bound for QW.

The set $\mathcal{F}_\theta^U$ is convex; hence for the purpose of minimizing a linear function one need only consider its extreme points. For the specific case of three phases and two space dimensions we shall give a complete classification of these extreme points in Section 4.

One would like to know whether our lower bound on QW is optimal, i.e. whether it is equal to QW. This amounts to asking whether $\mathcal{F}_\theta = \mathcal{F}_\theta^U$. We shall show that $\mathcal{F}_\theta$, too, is convex, so it is equivalent to ask whether each extreme point of $\mathcal{F}_\theta^U$ lies in $\mathcal{F}_\theta$. For three phases in two space dimensions the extreme points come in three types. The first two types are attained by sequentially laminated microstructures of rank 2; in particular, they are in $\mathcal{F}_\theta$. We have been unable to find any microstructure corresponding to an extreme point of type 3. The question whether $\mathcal{F}_\theta = \mathcal{F}_\theta^U$ is equivalent to whether the type 3 extreme points are attained by microstructures; this question, however, remains open.

The final section of this paper addresses the "three gradient problem." It asks whether three pairwise incompatible gradients can nevertheless be mutually compatible. In other words, given matrices a^1, a^2, and a^3 such that rank $(a^i - a^j) > 1$ for $i \neq j$, and defining W by (1.3), does $QW(\xi)$ vanish for some $\xi \notin \{a^1, a^2, a^3\}$? (We refer to Section 5 for a discussion motivating this question, and for a summary of related results.) Our analysis is restricted to two space dimensions; for maximum simplicitly it is focused primarily on the case when $\{a^i\}$ are simultaneously diagonal. Our main result is an equivalence between the three gradient problem and the attainability of our type 3 extreme points (see Theorem 5.4). It should be emphasized, however, that the three gradient problem has not been settled, since the attainability of type 3 extreme points remains open.

Our attention is focused throughout on $N \geq 3$ phases, because the situation for two phases is much simpler. That case is analyzed in [17,20,24]. When $N = 2$ it turns out that $\mathcal{F}_\theta = \mathcal{F}_\theta^U$, so the "Geometric Parameters Lower Bound" is actually a formula for QW.

It is natural to ask how our "Geometric Parameters Lower Bound" compares with the polyconvexification of W or the optimal translation bound. For two phases the Geometric Parameters bound can also be proved using the translation method [17,24]. For $N \geq 3$ phases we know no direct relation between geometric parameters and the translation method. Section 5 offers one indication of a possible connection: whenever we can prove that $QW(\xi) > 0$ for $\xi \notin \{a^1, a^2, a^3\}$ using polyconvexification, the same conclusion can also be deduced independently using geometric parameters (see Remark 5.6).

Acknowledgement. The tensor of geometric parameters that we study here was first considered a few years ago by G. Milton, in an attempt to improve upon the Haskin-Shtrikman bounds for the effective conductivity of a multicomponent composite. His analysis included the basic properties (3.2)-(3.6) and the layering formula (3.13)-(3.14), though it did not extend to a classification of the extreme points. We are grateful to Milton for sharing with us his insight and his unpublished notes on this subject.

2. Relaxation and the tensor of geometric parameters. This section defines the tensor of geometric parameters associated to a microstructure, and explains its relationship to the relaxed energy. We assume that W has the form (1.3) for some $\{a^i\}_{i=1}^N$. Each a^i is an $m \times n$ matrix; in other words we are considering $u : \mathbf{R}^n \to \mathbf{R}^m$.

The relaxed energy QW has already been defined by (1.2). However we plan to use Fourier analysis, so it is more convenient to use a characterization based on periodic functions. We choose $C = [0, 2\pi]^n$ as the period cell in $\mathbf{R}^n$, and we write $⨍ f$ for the average value of a C-periodic function f. Then QW has the alternative characterization

$$QW(\xi) = \inf_{\varphi \text{ per}} ⨍ W(\xi + \nabla\varphi), \tag{2.1}$$

where φ ranges over C-periodic H^1 functions from $\mathbf{R}^n$ to $\mathbf{R}^m$ [17].

Because of the special form of W, (2.1) can also be written another way:

$$QW(\xi) = \inf_{\varphi} \inf_{\chi_i} ⨍ \sum \chi_i |\xi + \nabla\varphi - a^i|^2, \tag{2.2}$$

where φ ranges over periodic functions as before, and $\{\chi_i\}_{i=1}^N$ range over C-periodic characteristic functions:

$$\begin{aligned} \chi_i(x) &= 0 \text{ or } 1 \text{ for every } x \\ \chi_i(x)\chi_j(x) &= 0 \text{ for } i \neq j \\ \sum_{i=1}^N \chi_i(x) &= 1 \text{ for every } x. \end{aligned}$$

The equivalence of (2.1) and (2.2) is elementary: if φ is fixed then an optimal choice of $\{\chi_i\}$ for (2.2) has

$$\chi_i(x) = 1 \Longrightarrow |\xi + \nabla\varphi - a^i|^2 = \min_{1\le j\le N} |\xi + \nabla\varphi - a^j|^2,$$

and substitution yields (2.1). We think of $\{\chi_i\}$ as a (spatially periodic) phase arrangement or *microstructure*. One sees from (2.2) that QW is obtained by minimizing the energy of the N-phase system over all possible microstructures, while holding the average gradient equal to ξ.

We now rewrite (2.2) by computing the minimum over φ explicitly, for given $\{\chi_i\}$. Writing

$$\theta_i = ⨍\chi_i \tag{2.4}$$

for the volume fraction of the ith phase, one easily sees that the integral in (2.2) is equal to

$$\sum_{i=1}^N \theta_i|\xi - a^i|^2 + ⨍\left(|\nabla\varphi|^2 - 2\langle\nabla\varphi, \sum \chi_i a^i\rangle\right). \tag{2.5}$$

(Here and henceforth we write $\langle A, B\rangle = tr(AB^T)$ for the inner product of $m\times n$ matrices, and $\nabla\varphi$ represents the $m \times n$ matrix $(\nabla\varphi)_{\alpha\beta} = \partial\varphi_\alpha/\partial x_\beta$.) We are fixing $\{\chi_i\}$, so the first term in (2.5) is determined. Hence the optimal φ solves

$$\inf_{\varphi \text{ per}} ⨍|\nabla\varphi|^2 - 2\langle\nabla\varphi, \sum\chi_i a^i\rangle, \tag{2.6}$$

or equivalently

$$\Delta\varphi_\alpha - \sum_{\beta=1}^n \frac{\partial}{\partial x_\beta}\left(\sum_{i=1}^N \chi_i a^i_{\alpha\beta}\right) = 0. \tag{2.7}$$

It is a straightforward matter to solve (2.7) by means of Fourier analysis. Writing

$$\chi_j(x) = \sum_{k\in\mathbf{Z}^n} \widehat{\chi}_j(k)e^{ik\cdot x}, \tag{2.8}$$

one finds after some calculation that

$$|\widehat{\nabla\varphi}|^2(k) = \sum \frac{k_\alpha k_\beta \widehat{\chi}_i(k)\overline{\widehat{\chi}_j(k)}a^i_{\gamma\alpha}a^j_{\gamma\beta}}{|k|^2} \tag{2.9}$$

for any $k \in \mathbf{Z}^n$, $k \neq 0$. (The sum in (2.9) is over all repeated indices: α and β range from 1 to n, i and j range from 1 to N, and γ runs from 1 to m.) Now, an integration by parts based on (2.7) gives

$$⨍|\nabla\varphi|^2 - 2\langle\nabla\varphi, \sum\chi_i a^i\rangle = -⨍|\nabla\varphi|^2.$$

Using this along with (2.9) and Plancherel's Theorem, we deduce that

$$\inf(2.6) = -\sum_{k\neq 0}\sum_{\alpha,\beta,\gamma,i,j} \frac{k_\alpha k_\beta}{|k|^2}\widehat{\chi}_i(k)\overline{\widehat{\chi}_j(k)}a^i_{\gamma\alpha}a^j_{\gamma\beta}.$$

We have thus derived the following alternative representation of (2.2):

$$QW(\xi) = \inf_{\chi_i}\left\{\sum \theta_i|\xi - a^i|^2 - \sum \frac{k_\alpha k_\beta}{|k|^2}\widehat{\chi}_i(k)\overline{\widehat{\chi}_j(k)}a^i_{\gamma\alpha}a^j_{\gamma\beta}\right\}. \tag{2.10}$$

The right hand side of (2.10) depends only on certain features of the microstructure $\{\chi_i\}$. To clarify this dependence, we define the *tensor of geometric parameters* associated to $\{\chi_i\}$ by

$$F_{ij\alpha\beta} = \sum_{k\neq 0} \frac{k_\alpha k_\beta}{|k|^2}\widehat{\chi}_i(k)\overline{\widehat{\chi}_j(k)}. \tag{2.11}$$

(It is not a true tensor, since i and j refer to phases rather than spatial dimensions.) Let V be the simplex of all possible volume fractions:

$$V = \left\{\theta = (\theta_1, \ldots, \theta_N) : \theta_i \geq 0, \sum_{i=1}^{N}\theta_i = 1\right\}. \tag{2.12}$$

For any $\theta \in V$, let $\mathcal{F}_\theta$ be the set of all geometric parameters attainable with volume fraction θ:

$$\begin{aligned}\mathcal{F}_\theta = \;& \text{The closure of the set of all } F_{ij\alpha\beta} \\ & \text{defined by (2.11), as } \{\chi_i\} \text{ ranges over} \\ & C\text{-periodic characteristic functions with} \\ & \textstyle\fint \chi_i = \theta_i.\end{aligned} \tag{2.13}$$

Then (2.10) may be expressed as

$$QW(\xi) = \inf_{\theta\in V}\inf_{F\in\mathcal{F}_\theta}\left\{\sum \theta_i|\xi - a^i|^2 - \sum F_{ij\alpha\beta}a^i_{\gamma\alpha}a^j_{\gamma\beta}\right\}. \tag{2.14}$$

We emphasize that (2.14) is an exact "formula" for $QW(\xi)$. It is better than the other representations (1.2) or (2.1) or (2.2), because it involves only a finite dimensional minimization. It is not a computable representation, however, because we do not know the exact form of the set $\mathcal{F}_\theta$. Our "Geometric Parameters Lower Bound" will be obtained in Section 3 by replacing $\mathcal{F}_\theta$ with a (possibly) larger set $\mathcal{F}^U_\theta$ which is known explicitly.

To place the tensor of geometric parameters in its proper mathematical context, we now explain its relationship to the H-measure of the microstructure [29], also called its microlocal defect measure [13]. Given a (spatially periodic) microstructure

$\{\chi_i\}$, the associated H-measure is the symmetric matrix-valued measure on S^{n-1} defined by

$$\mu_{ij} = Re \sum_{k \neq 0} \widehat{\chi}_i(k) \overline{\widehat{\chi}_j(k)} \delta_{\frac{k}{|k|}}, \tag{2.15}$$

where $\delta_{\frac{k}{|k|}}$ is the Dirac measure concentrated at $\frac{k}{|k|}$. The tensor of geometric parameters keeps track precisely of the second moments of the H-measure. Indeed, it is an easy consequence of the definitions (2.11) and (2.15) that

$$F_{ij\alpha\beta} = \int_{S^{n-1}} \eta_\alpha \eta_\beta d\mu_{ij}(\eta). \tag{2.16}$$

A discussion of multiwell energies based on H-measures rather than geometric parameters will be found in [17].

3. Properties of the geometric parameters. This section presents the known algebraic properties of the tensor of geometric parameters. It also discusses the geometric parameters associated with sequentially laminated microstructures. Its main result is the "Geometric Parameters Lower Bound" for QW, (3.8).

We introduce some notation for use in the following proposition: given any $\theta \in V$, let

$$\Gamma_{ij} = \begin{cases} \theta_i(1-\theta_i) & \text{if } i = j \\ -\theta_i\theta_j & \text{if } i \neq j. \end{cases} \tag{3.1}$$

PROPOSITION 3.1. *If $F \in \mathfrak{F}_\theta$, then*

$$F_{ij\alpha\beta} = F_{ji\alpha\beta} = F_{ij\beta\alpha}, \tag{3.2}$$

$$\sum_{i=1}^N F_{ij\alpha\beta} = \sum_{j=1}^N F_{ij\alpha\beta} = 0, \tag{3.3}$$

$$\sum_{\alpha=1}^n F_{ij\alpha\alpha} = \Gamma_{ij}, \tag{3.4}$$

$$\sum F_{ij\alpha\beta}\, q_{i\alpha} q_{j\beta} \geq 0 \textit{ for any real } N \times n \textit{ matrix } q_{i\alpha}, \tag{3.5}$$

$$\sum (\delta_{\alpha\beta}\Gamma_{ij} - F_{ij\alpha\beta})\, q_{i\alpha} q_{j\beta} \geq 0 \textit{ for any real } N \times n \textit{ matrix } q_{i\alpha}. \tag{3.6}$$

Proof. The symmetries (3.2) are immediate from the definition (2.11), using the fact that $\{\chi_i\}$ are real. Property (3.3) follows from the fact that $\sum_{i=1}^N \chi_i = 1$. To prove (3.4) we use Plancherel's Theorem:

$$\begin{aligned} \sum_{\alpha=1}^n F_{ij\alpha\alpha} &= \sum_{k \neq 0} \widehat{\chi}_i(k) \overline{\widehat{\chi}_j(k)} \\ &= \fint (\chi_i - \theta_i)(\chi_j - \theta_j) = \Gamma_{ij}. \end{aligned}$$

For (3.5) we observe that

$$\sum F_{ij\alpha\beta}q_{i\alpha}q_{j\beta} = \sum_{k\neq 0}\sum \widehat{\chi}_i(k)\frac{k_\alpha}{|k|}q_{i\alpha}\overline{\widehat{\chi}_j(k)\frac{k_\beta}{|k|}q_{j\beta}},$$

which is a sum of squares. Property (3.6) follows similarly from the observation that

$$\sum_{k\neq 0}\sum \widehat{\chi}_i(k)\left(\delta_{\alpha\gamma}-\frac{k_\alpha k_\gamma}{|k|^2}\right)q_{i\alpha}\overline{\widehat{\chi}_j(k)\left(\delta_{\beta\gamma}-\frac{k_\beta k_\gamma}{|k|^2}\right)q_{j\beta}}$$

is a sum of squares. □

In view of Proposition 3.1, $\mathcal{F}_\theta$ is contained in the set

$$\mathcal{F}_\theta^U = \{F = (F_{ij\alpha\beta}) : F \text{ satisfies (3.2)-(3.6) }\}. \tag{3.7}$$

We therefore deduce

COROLLARY 3.2. *(The "Geometric Parameters Lower Bound.") For W of the form (1.3),*

$$QW(\xi) \geq \inf_{\theta\in V}\inf_{F\in\mathcal{F}_\theta^U}\left\{\sum\theta_i|\xi-a^i|^2 - \sum F_{ij\alpha\beta}a^i_{\gamma\alpha}a^j_{\gamma\beta}\right\}. \tag{3.8}$$

The advantage of (3.8) is that it involves minimization over a set which is known explicitly. Thus (3.8) is a computable lower bound, at least in principle. It is not easy to evaluate, however, because $\mathcal{F}_\theta^U$ is defined by inequalities (3.5)-(3.6) as well as linear relations (3.2)-(3.4). We shall make it easier to evaluate in Section 4, for the special case of 3 wells in two space dimensions.

It is interesting to observe that in two space dimensions properties (3.5) and (3.6) are equivalent. This is analogous to the fact that for 2×2 symmetric matrices A, $A\geq 0 \iff (trA)I - A \geq 0$ in the sense of quadratic forms.

LEMMA 3.3. *In space dimension $n=2$, suppose that F satisfies (3.2) and (3.4). Then it satisfies (3.5) iff it satisfies (3.6).*

Proof. Consider the $2N\times 2N$ matrix

$$G = \begin{pmatrix} F_{11} & \dots & F_{1N} \\ \vdots & \ddots & \vdots \\ F_{N1} & \dots & F_{NN}\end{pmatrix},$$

in which F_{ij} represents the 2×2 block $(F_{ij\alpha\beta})_{\alpha,\beta=1,2}$. By (3.2), G is symmetric. Property (3.5) is equivalent to the statement that $G\geq 0$. This holds for G iff it holds for any matrix similar to G. Let $R = \begin{pmatrix} 0 & 1 \\ -1 & 0\end{pmatrix}$, and consider the $2N\times 2N$ rotation

$$M = \begin{pmatrix} R & & 0 \\ & \ddots & \\ 0 & & R\end{pmatrix}.$$

Noting that $R^TAR = (trA)I - A$ for any symmetric 2×2 matrix, and making use of (3.4), we find that

$$M^TGM = \begin{pmatrix} \Gamma_{11}I - F_{11} & \dots & \Gamma_{1N}I - F_{1N} \\ \vdots & \ddots & \vdots \\ \Gamma_{N1}I - F_{N1} & \dots & \Gamma_{NN}I - F_{NN} \end{pmatrix}.$$

The positivity of M^TGM is equivalent to (3.6). □

We deduce as a consequence

COROLLARY 3.4. *In space dimension two, the set* $\mathcal{F}^U_\theta$ *is invariant under the map*

$$F_{ij\alpha\beta} \to \Gamma_{ij}\delta_{\alpha\beta} - F_{ij\alpha\beta}.$$

It is well-known that the convexification of W always provides a lower bound for QW. As a partial confirmation that (3.8) is a good bound, we now verify that it always lies above the convexification.

LEMMA 3.5. *Let* CW *denote the convexification of* W*. Then*

$$\sum \theta_i|\xi - a^i|^2 - \sum F_{ij\alpha\beta}a^i_{\gamma\alpha}a^j_{\gamma\beta} \geq CW(\xi) \tag{3.9}$$

for any $\theta \in V$ *and any* $F \in \mathcal{F}^U_\theta$*. In particular, the "geometric parameters lower bound" is greater than or equal to* CW*.*

Proof. First let us show that

$$CW(\xi) = \inf_{\theta \in V} |\xi - \sum \theta_i a^i|^2. \tag{3.10}$$

Indeed, for any W one has

$$CW(\xi) = \inf_{\int \lambda d\mu(\lambda) = \xi} \int W(\lambda)d\mu(\lambda),$$

in which μ ranges over probability measures on the space of $m \times n$ matrices, see e.g. [9]. Since each "well" $|\xi - a^i|^2$ is convex, an easy application of Jensen's inequality shows that the optimal μ will be a sum of N point masses. It follows that

$$CW(\xi) = \inf_{\sum \theta_i\xi_i = \xi} \sum \theta_i|\xi_i - a^i|^2,$$

where $\theta = (\theta_1, \dots, \theta_N)$ ranges over V and ξ_i over $m \times n$ matrices. Optimization over ξ_i (with θ held fixed) yields (3.10).

Now let us rearrange the expression in (3.10):

$$\begin{aligned} |\xi - \sum \theta_i a^i|^2 &= |\xi|^2 - 2\langle \xi, \sum_i \theta_i a^i \rangle + \sum_{i,j} \theta_i\theta_j \langle a^i, a^j \rangle \\ &= \sum_i \theta_i|\xi - a^i|^2 - \sum_i \theta_i|a^i|^2 + \sum_{i,j} \theta_i\theta_j\langle a^i, a^j \rangle \\ &= \sum \theta_i|\xi - a^i|^2 - \sum \Gamma_{ij}\delta_{\alpha\beta}\, a^i_{\gamma\alpha}a^j_{\gamma\beta}. \end{aligned} \tag{3.11}$$

We use (3.6) and (3.11) to bound the left hand side of (3.9):

$$\begin{aligned}\sum \theta_i|\xi - a^i|^2 - \sum F_{ij\alpha\beta}a^i_{\gamma\alpha}a^j_{\gamma\beta} &\geq \sum \theta_i|\xi - a^i|^2 - \sum \Gamma_{ij}\delta_{\alpha\beta}a^i_{\gamma\alpha}a^j_{\gamma\beta} \\ &= |\xi - \sum \theta_i a^i|^2 \\ &\geq CW(\xi),\end{aligned} \tag{3.12}$$

as asserted. □

A crucial question is whether or not $\mathfrak{F}_\theta = \mathfrak{F}^U_\theta$. This remains open. An affirmative answer would provide, for each $F \in \mathfrak{F}^U_\theta$, a microstructure whose tensor of geometric parameters is equal to F. To make progress in this direction, it is important to look for microstructures whose tensors of geometric parameters are explicitly computable. One such class are the microstructures obtained by *sequential lamination*. That construction has played a central role in recent work on the effective moduli of composites, see e.g. [2,22,28]. In the present context the construction is as follows: consider any pair of microstructures, with possibly different volume fractions $\theta', \theta'' \in V$, and with tensors of geometric parameters $F' \in \mathfrak{F}_{\theta'}$, $F'' \in \mathfrak{F}_{\theta''}$. We construct a new microstructure by layering the two with one another, using volume fractions ρ and $1-\rho$ respectively, in layers orthogonal to some unit vector k. If the length scale of the microstructure is small compared to the length scale of the layering, then the geometric parameters of the new microstructure depend only on F', F'', θ', θ'', ρ, and k:

PROPOSITION 3.6. *Suppose that $F' \in \mathfrak{F}_{\theta'}$, $F'' \in \mathfrak{F}_{\theta''}$. For any ρ, $0 < \rho < 1$, and any unit vector k, consider the microstructure associated with "layering F' and F'' in volume fractions ρ, $1-\rho$ with layers orthogonal to k," as described in more detail above. It has volume fractions*

$$\theta_i = \rho\theta'_i + (1-\rho)\theta''_i \tag{3.13}$$

and its tensor of geometric parameters is

$$F_{ij\alpha\beta} = \rho F'_{ij\alpha\beta} + (1-\rho)F''_{ij\alpha\beta} + \rho(1-\rho)(\theta'_i - \theta''_i)(\theta'_j - \theta''_j)k_\alpha k_\beta. \tag{3.14}$$

Proof. Assertion (3.13) is elementary. Assertion (3.14) follows by (2.16) from the analogous layering formula for H-measures, see formula (8.16) of [17]. □

An important corollary of the layering formula is the convexity of $\mathfrak{F}_\theta$:

COROLLARY 3.7. *For any $\theta \in V$, $\mathfrak{F}_\theta$ is convex.*

Proof. Let $F', F'' \in \mathfrak{F}_\theta$. Then (since both have the *same* volume fractions) (3.13)-(3.14) give

$$F = \rho F' + (1-\rho)F'' \in \mathfrak{F}_\theta$$

for any ρ,$0 < \rho < 1$. Notice that in this case the result of layering is independent of the direction of the layers. □

The term "sequential lamination" arises from the observation that the layering process can be repeated any numbers of times. We begin with the N "pure" phases, described by $F \equiv 0$ and $\theta = (0, \ldots, 1, \ldots, 0)$. (The 1 is in the ith place for the ith phase.) One says a microstructure is "sequentially laminated of rank r" if it is obtained from these by r applications of the layering construction. We denote by $\mathfrak{F}_\theta^L$ the set of all geometric parameters attainable by sequential lamination. Since the layering formula generally mixes microstructures with different volume fractions, $\mathfrak{F}_\theta^L$ is best described as the slice at $\theta \in V$ of a set in the product space of $\{$ volume fractions $\} \times \{$ geometric parameters $\}$:

$$\mathfrak{F}_\theta^L = \begin{array}{l} \text{slice at } \theta \in V \text{ of the smallest closed} \\ \text{set of pairs } (\theta, F) \text{ that contains the} \\ N \text{ pure phases and is preserved} \\ \text{under application of the layering} \\ \text{procedure (3.13) - (3.14).} \end{array} \tag{3.15}$$

We do not have an explicit representation of $\mathfrak{F}_\theta^L$.

Since $\mathfrak{F}_\theta^L \subset \mathfrak{F}_\theta$, we could formulate an associated upper bound for QW:

$$QW(\xi) \leq \inf_{\theta \in V} \inf_{F \in \mathfrak{F}_\theta^L} \left\{ \sum \theta_i |\xi - a^i|^2 - \sum F_{ij\alpha\beta} a^i_{\gamma\alpha} a^j_{\gamma\beta} \right\}. \tag{3.16}$$

This is not a new bound, however: the right side of (3.16) is precisely the rank-one convexification of W. Indeed, for any W the rank-one convexification can be characterized as the best upper bound achievable by a sequentially laminated microstructure [8,18,23]. The right hand side of (3.16) is the same thing, specialized to W of the form (1.3) and phrased in terms of geometric parameters.

To recapitulate, the main goal is to find $\mathfrak{F}_\theta$. We have described sets $\mathfrak{F}_\theta^L$ and $\mathfrak{F}_\theta^U$ such that

$$\mathfrak{F}_\theta^L \subseteq \mathfrak{F}_\theta \subseteq \mathfrak{F}_\theta^U.$$

We shall show in Section 5 that the inclusion $\mathfrak{F}_\theta^L \subset \mathfrak{F}_\theta^U$ is strict, for some choices of $\theta \in V$, in the context of three wells and two space dimensions. It remains a possibility that $\mathfrak{F}_\theta^L = \mathfrak{F}_\theta$; in this case the list of properties (3.2) - (3.6) is incomplete. It is also a possibility that $\mathfrak{F}_\theta = \mathfrak{F}_\theta^U$; in that case the class of sequentially laminated microstructures is not rich enough to include an energetically optimal configuration. Finally, it might be the case that both inequalities are strict. The question whether $\mathfrak{F}_\theta^L = \mathfrak{F}_\theta$ corresponds to asking whether, for W of the form (1.3), the quasiconvexification equals the rank-one convexification. The question whether $\mathfrak{F}_\theta^U = \mathfrak{F}_\theta$ is analogous to (but apparently different from) asking whether, for such W, the quasiconvexification equals the polyconvexification.

4. Extreme points for three wells in two dimensions. This section determines the extreme points of the convex set $\mathcal{F}_\theta^U$ in the special case of three wells and two space dimensions. The restriction $N = 3$, $n = 2$ applies through the section.

We have already observed in Lemma 3.3 that it is useful to view a tensor of geometric parameters as a block matrix:

$$F = \begin{pmatrix} F_{11} & F_{12} & F_{13} \\ F_{21} & F_{22} & F_{23} \\ F_{31} & F_{32} & F_{33} \end{pmatrix} \tag{4.1}$$

where F_{ij} represents the 2×2 block $F_{ij\alpha\beta}$. By (3.2)-(3.3) each block is symmetric, and there are only three independent blocks:

$$F = \begin{pmatrix} A & B & -A-B \\ B & C & -B-C \\ -A-B & -B-C & A+2B+C \end{pmatrix} \tag{4.2}$$

with $A = F_{11}$, $B = F_{12}$, $C = F_{22}$. Condition (3.4) is equivalent to

$$tr A = \theta_1(1-\theta_1), \quad tr B = -\theta_1\theta_2, \quad tr C = \theta_2(1-\theta_2). \tag{4.3}$$

The following proposition gives conditions on A, B, and C which are equivalent to F satisfying (3.2) - (3.6).

PROPOSITION 4.1. *For $N = 3$, $n = 2$, the set $\mathcal{F}_\theta^U$ is in $1-1$ correspondence with the triplets of symmetric, 2×2 matrices (A, B, C) that satisfy the trace relations (4.3) and the following positivity condition:*

$$\langle Ax, x\rangle + 2\langle Bx, y\rangle + \langle Cy, y\rangle \geq 0 \quad \text{for all } x, y \in \mathbb{R}^2. \tag{4.4}$$

Proof. We have only to prove that (4.4) is equivalent to (3.5) - (3.6). By Lemma 3.3, (3.5) and (3.6) are equivalent, so it suffices to consider (3.5). Written in terms of A, B, and C, this condition asserts that

$$\begin{aligned} \langle Ax, x\rangle + &\langle By, y\rangle + \langle (A+2B+C)z, z\rangle \\ &+2\langle Bx, y\rangle - 2\langle (A+B)x, z\rangle - 2\langle (B+C)y, z\rangle \geq 0 \end{aligned} \tag{4.5}$$

for every $x, y, z \in \mathbb{R}^2$. Clearly (4.5) implies (4.4) by taking $z = 0$. On the other hand, (4.5) can also be written as

$$\langle A(x-z), x-z\rangle + 2\langle B(x-z), y-z\rangle + \langle C(y-z), y-z\rangle \geq 0.$$

This condition is implied by (4.4). $\square$

Let us examine the positivity condition (4.4) more closely. If A is invertible then (4.4) holds if and only if

$$A \geq 0, C \geq 0, \text{ and } C - BA^{-1}B \geq 0 \tag{4.6}$$

in the sense of quadratic forms. (Actually, the condition $C \geq 0$ in (4.6) is redundant, since $BA^{-1}B \geq 0$.) Indeed, it is obvious that (4.4) implies $A \geq 0$ and $C \geq 0$. If we rewrite (4.4) as

$$|A^{1/2}x + A^{-1/2}By|^2 + \langle (C - BA^{-1}B)y, y\rangle \geq 0, \tag{4.7}$$

then the positivity of the Schur complement $C - BA^{-1}B$ becomes clear as well. Conversely, (4.6) clearly implies (4.7) and hence also (4.4).

When A is not invertible, we can retain the equivalence of (4.4) and (4.6) by interpreting the latter properly. Specifically, we interpret "$C - BA^{-1}B \geq 0$" as the assertion that

$$\lim_{\varepsilon \downarrow 0} C - B(A + \varepsilon I)^{-1}B \geq 0. \tag{4.8}$$

Since (4.4) is a closed condition, one easily checks that (4.6) (with the convention (4.8)) is equivalent to (4.4) even when A is not invertible.

Condition (4.8) really amounts to two separate assertions:

(4.9) The range of B is contained in the range of A; and

(4.10) $C - BA^+B \geq 0$,

where A^+ is the Moore-Penrose inverse of A, i.e. the inverse of the operator obtained by restricting A to the orthogonal complement of its kernel (which is also its range, since A is symmetric). In our 2×2 setting A fails to be invertible only if $A = 0$ or if A has rank $1 : A = \alpha k \otimes k$, $\alpha > 0$, $|k| = 1$. In the former case (4.9) forces $B = 0$; in the latter case it forces $B = \beta k \otimes k$, and (4.10) becomes $C - (\beta^2/\alpha) k \otimes k \geq 0$. We note that when A is not invertible, the quadratic form (4.4) can still be written in the form (4.7) provided that (4.8) holds; here "$A^{-1/2}$" must be interpreted as the nonnegative, symmetric square root of A^+.

All the above remarks apply just as well when the roles of A and C are interchanged. In particular, (4.4) is also equivalent to

$$A \geq 0, \quad C \geq 0, \quad A - BC^{-1}B \geq 0, \tag{4.11}$$

with the convention analogous to (4.8) when C is not invertible.

Our classification of the extreme points will make use of the following well-known fact.

LEMMA 4.2. *Let M_+ denote the set of nonnegative, symmetric, $n \times n$ matrices with trace 1. Then the extreme points of M_+ are precisely the rank-one matrices $k \otimes k$ with $|k| = 1$.*

Proof. To see that $k \otimes k$ is extreme, suppose $k \otimes k = \rho M_1 + (1 - \rho) M_2$ with $M_1, M_2 \in \mathsf{M}_+$. Then $\langle M_1 v, v\rangle = \langle M_2 v, v\rangle = 0$ for any $v \perp k$, and it follows that $M_1 = M_2 = k \otimes k$. There can be no other extreme points, because the spectral decomposition of any $M \in \mathsf{M}_+$ expresses it as a convex combination of rank-one matrices. □

We are now ready to classify the extreme points of $\mathcal{F}_\theta^U$.

THEOREM 4.3. *Let (A,B,C) be a triplet of matrices corresponding to a point of $\mathcal{F}^U_\theta$. They correspond to an extreme point precisely if they fall in one of the following three classes.*

- *Extreme points of type 1:*

$$A = \theta_1(1-\theta_1)k\otimes k, \quad B = -\theta_1\theta_2 k\otimes k, \quad C = \theta_2(1-\theta_2)k\otimes k \tag{4.12}$$

for some unit vector k.

- *Extreme points of type 2: either*

$$\begin{aligned} A &= \theta_1(1-\theta_1)k\otimes k, \quad B = -\theta_1\theta_2 k\otimes k \\ C &= \frac{\theta_1^2\theta_2^2}{\theta_1(1-\theta_1)}k\otimes k + \left[\theta_2(1-\theta_2) - \frac{\theta_1^2\theta_2^2}{\theta_1(1-\theta_1)}\right] l\otimes l \end{aligned} \tag{4.13}$$

or else

$$\begin{aligned} C &= \theta_2(1-\theta_2)k\otimes k, \quad B = -\theta_1\theta_2 k\otimes k \\ A &= \frac{\theta_1^2\theta_2^2}{\theta_2(1-\theta_2)}k\otimes k + \left[\theta_1(1-\theta_1) - \frac{\theta_1^2\theta_2^2}{\theta_2(1-\theta_2)}\right] l\otimes l \end{aligned} \tag{4.14}$$

for some unit vectors k and l.

- *Extreme points of type 3: A and B are both invertible, $A^{-1}B \neq \lambda I$ for $\lambda \in \mathbf{R}$, and $C = BA^{-1}B$.*

Proof. Suppose first that $A=0$. Then the trace relations (4.3) force $\theta_1 = 0$ or 1 and $B=0$. The trace relations (4.3) and the positivity condition (4.6) permit any C with $C\geq 0$ and $trC = \theta_2(1-\theta_2)$. By Lemma 4.2, the only such extreme point is

$$A = 0, \quad B = 0, \quad C = \theta_2(1-\theta_2)k\otimes k$$

for some $|k|=1$. (It is simultaneously of types 1 and 2 due to the degeneracy $\theta_1 = 0$ or $\theta_1 = 1$.)

Next, suppose that A has rank one: $A = \theta_1(1-\theta_1)k\otimes k$ with $|k|=1$. By (4.3) and (4.6), B and C must satisfy $B = -\theta_1\theta_2 k\otimes k$ and

$$C - \frac{\theta_1^2\theta_2^2}{\theta_1(1-\theta_1)}k\otimes k \geq 0. \tag{4.15}$$

If the left side of (4.15) is strictly positive definite then C can be expressed as a convex combination of matrices satisfying both (4.15) and the trace condition. So if (A,B,C) is extreme then the left side of (4.15) equals zero or has rank one. These two alternatives correspond to (4.12) and (4.13) respectively. Let us check that such points are indeed extreme. If (A,B,C) are as in (4.12), consider a convex combination

$$(A,B,C) = \rho(A_1,B_1,C_1) + (1-\rho)(A_2,B_2,C_2).$$

Since the volume fraction θ is fixed, Lemma 4.2 yields $A_1 = A_2 = A$ and $C_1 = C_2 = C$. Since A has rank one, we also have $B_1 = B_2 = B$. Thus points of the form

(4.12) are extreme. Similarly, suppose that (A, B, C) are as in (4.13) and consider a convex combination as above. Lemma 4.2 still yields $A_1 = A_2 = A$, and it still follows that $B_1 = B_2 = B$. So C_1 and C_2 must satisfy the analogue of (4.15). An application of Lemma 4.2 shows that $C_1 = C_2 = C$, since the left hand side of (4.15) has rank one. Thus points of the form (4.13) are extreme.

If A is nonsingular but C is singular then we can repeat the preceding arguments with A replaced by C. This leads to either (4.12) or (4.14).

Now suppose that A and C are both nonsingular. If in addition $C - BA^{-1}B > 0$ then (A, B, C) lies in the interior of the region determined by the positivity condition (4.6); it is easy to see that such (A, B, C) are not extreme points. If $C - BA^{-1}B$ has rank one, then we see that (A, B, C) is not an extreme point as follows. Let $y_0 \neq 0$ satisfy $(C - BA^{-1}B)y_0 = 0$ and set $x_0 = -A^{-1}By_0$. Next, choose symmetric matrices A_0, B_0, and C_0 such that

$$\begin{gathered}\langle A_0x_0, x_0\rangle = 0, \quad \langle B_0x_0, y_0\rangle = 0, \quad \langle C_0y_0, y_0\rangle = 0 \\ tr A_0 = tr B_0 = tr C_0 = 0.\end{gathered} \tag{4.16}$$

Let $Q(x, y)$ denote the quadratic form in (4.4):

$$Q(x, y) = \langle Ax, x\rangle + 2\langle Bx, y\rangle + \langle Cy, y\rangle. \tag{4.17}$$

Our special choice of A_0, B_0, C_0 assures that

$$Q(x, y) \geq c\,\{|\langle A_0x, x\rangle| + |\langle B_0x, y\rangle| + |\langle C_0y, y\rangle|\} \tag{4.18}$$

for all $x, y \in \mathbf{R}^2$ if the constant $c > 0$ is chosen sufficiently small. Indeed, if (4.18) were to fail for every $c > 0$ then there would exist $x_1, y_1 \in \mathbf{R}^2$ such that $Q(x_1, y_1) = 0$ and

$$|\langle A_0x_1, x_1\rangle| + |\langle B_0x_1, y_1\rangle| + |\langle C_0y_1, y_1\rangle| = 1. \tag{4.19}$$

¿From (4.7) we see that $Q(x_1, y_1) = 0$ implies $x_1 = \lambda x_0$ and $y_1 = \lambda y_0$ for some $\lambda \in \mathbf{R}$, so (4.19) contradicts (4.16). Thus (4.18) holds for some $c > 0$. It follows that $(A \pm \varepsilon A_0, B \pm \varepsilon B_0, C \pm \varepsilon C_0)$ satisfies (4.6) when ε is sufficiently small. So

$$(A, B, C) = \frac{1}{2}(A + \varepsilon A_0,\ B + \varepsilon B_0,\ C + \varepsilon C_0) + \frac{1}{2}(A - \varepsilon A_0,\ B - \varepsilon B_0,\ C - \varepsilon C_0)$$

is not an extreme point.

We have shown that for A and C invertible, the only possibility of an extreme point is when $C = BA^{-1}B$. If it should happen that $A^{-1}B = \lambda I$, then this is not an extreme point. Indeed, if $B = \lambda A$ and $C = BA^{-1}B = \lambda^2 A$, then the positivity condition (4.4) becomes

$$\langle A(x + \lambda y), x + \lambda y\rangle \geq 0,$$

which holds whenever A is positive definite. Choosing any matrix δA such that $tr\delta A = 0$ and $A + \delta A \geq 0$, we have

$$(A, \lambda A, \lambda^2 A) = \frac{1}{2}(A + \delta A,\ \lambda(A + \delta A),\ \lambda^2(A + \delta A)) + \frac{1}{2}(A - \delta A,\ \lambda(A - \delta A),\ \lambda^2(A - \delta A)),$$

so $(A, \lambda A, \lambda^2 A)$ is not an extreme point. We note that this exceptional case can occur only for certain θ: the trace relations (4.3) force $\lambda = -\theta_2/(1 - \theta_1)$ and $\theta_1\theta_2 = (1 - \theta_1)(1 - \theta_2)$.

The only remaining task is to show that our "extreme points of type 3" are indeed extreme. Consider a convex combination

$$(A, B, C) = \rho(A_1, B_1, C_1) + (1 - \rho)(A_2, B_2, C_2).$$

Since A, B, and C are nonsingular, we may suppose without loss of generality that A_i, B_i, and C_i are also nonsingular. As a first step we shall show that $C_i = B_i A_i^{-1} B_i$ and $A_i^{-1} B_i = A^{-1} B$ for $i = 1, 2$. Indeed, consider the quadratic from $Q(x, y)$ associated to (A, B, C), defined by (4.17), and the analogous forms $Q_i(x, y)$ associated to (A_i, B_i, C_i). Since $C = BA^{-1}B$,

$$Q(x, y) = |A^{1/2}x + A^{-1/2}By|^2,$$

which vanishes on the subspace $x = -A^{-1}By$. Since each Q_i is nonnegative, Q_1 and Q_2 must vanish on this subspace. We have

$$Q_1(x, y) = |A_1^{1/2}x + A_1^{-1/2}By|^2 + \langle (C_1 - B_1 A_1^{-1} B_1)y, y\rangle.$$

As y varies over R^2 with $x = -A^{-1}By$ this forces $C_1 = B_1 A_1^{-1} B_1$ and $A_1^{-1} B_1 = A^{-1}B$. The same argument shows that $C_2 = B_2 A_2^{-1} B_2$ and $A_2^{-1} B_2 = A^{-1}B$.

Let $e_i, i = 1, 2$, be the generalized eigenvectors

$$Be_1 = -\lambda_1 Ae_1$$
$$Be_2 = -\lambda_2 Ae_2$$

with $\langle Ae_1, e_1\rangle = \langle Ae_2, e_2\rangle = 1$ and $\langle Ae_1, e_2\rangle = 0$. Notice that $\lambda_1 \neq \lambda_2$, since otherwise we would have $A^{-1}B = -\lambda I$. From the preceding paragraph we deduce that

$$B_i e_1 = -\lambda_1 A_i e_1$$
$$B_i e_2 = -\lambda_2 A_i e_2$$

for $i = 1, 2$. Since the eigenvalues are distinct, this forces

$$\langle B_i e_1, e_2\rangle = 0, \quad \langle A_i e_1, e_2\rangle = 0,$$

using the symmetry of A_i and B_i.

We return now to the quadratic forms $Q(x,y)$, $Q_1(x,y)$, and $Q_2(x,y)$. Consider the associated forms on R^2 defined by

$$\widetilde{Q}(s,t) = Q(se_1, te_1) = \begin{pmatrix} s \\ t \end{pmatrix}^T \begin{pmatrix} \langle Ae_1, e_1\rangle & \langle Be_1, e_1\rangle \\ \langle Be_1, e_1\rangle & \langle BA^{-1}Be_1, e_1\rangle \end{pmatrix} \begin{pmatrix} s \\ t \end{pmatrix},$$

and similarly for $\widetilde{Q}_1$, $\widetilde{Q}_2$. By the preceding paragraph, all three forms vanish on the one-dimensional subspace $s = \lambda_1 t$. By hypothesis they are all nonnegative, and

$$\widetilde{Q} = \rho\widetilde{Q}_1 + (1-\rho)\widetilde{Q}_2.$$

It follows that the matrices associated to $\widetilde{Q}_1$ and $\widetilde{Q}_2$ are multiples of that associated to $\widetilde{Q}$. Thus

$$\begin{aligned} \begin{pmatrix} \langle A_1e_1, e_1\rangle & \langle B_1e_1, e_1\rangle \\ \langle B_1e_1, e_1\rangle & \langle B_1A_1^{-1}B_1e_1, e_1\rangle \end{pmatrix} &= \mu_1 \begin{pmatrix} \langle Ae_1, e_1\rangle & \langle Be_1, e_1\rangle \\ \langle Be_1, e_1\rangle & \langle BA^{-1}Be_1, e_1\rangle \end{pmatrix} \\ &= \mu_1 \begin{pmatrix} 1 & -\lambda_1 \\ -\lambda_1 & \lambda_1^2 \end{pmatrix}. \end{aligned}$$

A similar relation holds for A_2 and B_2, with a different constant μ_2. Summarizing this information, we have:

$$\begin{aligned} \langle A_1e_1, e_1\rangle &= \mu_1, & \langle B_1e_1, e_1\rangle &= -\mu_1\lambda_1, \\ \langle A_2e_1, e_1\rangle &= \mu_2, & \langle B_2e_1, e_1\rangle &= -\mu_2\lambda_1, \end{aligned}$$

A parallel argument using e_2 in place of e_1 gives

$$\begin{aligned} \langle A_1e_2, e_2\rangle &= \nu_1, & \langle B_1e_2, e_2\rangle &= -\nu_1\lambda_2 \\ \langle A_2e_2, e_2\rangle &= \nu_2, & \langle B_2e_2, e_2\rangle &= -\nu_2\lambda_2, \end{aligned}$$

for two new constants ν_1, ν_2.

We claim that the trace relations force $\mu_1 = \mu_2 = \nu_1 = \nu_2 = 1$. Indeed, since e_1 and e_2 span $\mathbf{R}^2$, $e_1 \otimes e_1$, $e_2 \otimes e_2$, and $(e_1 \otimes e_2 + e_2 \otimes e_1)/2$ span the space of symmetric 2×2 matrices. So there are constants α, β, and γ such that

$$I = \alpha e_1 \otimes e_1 + \beta e_2 \otimes e_2 + \frac{\gamma}{2}(e_1 \otimes e_2 + e_2 \otimes e_1).$$

The above relations give

$$\begin{aligned} tr A &= \alpha + \beta & tr B &= -(\lambda_1\alpha + \lambda_2\beta) \\ tr A_1 &= \alpha\mu_1 + \beta\nu_1 & tr B_1 &= -(\lambda_1\alpha\mu_1 + \lambda_2\beta\nu_1) \\ tr A_2 &= \alpha\mu_2 + \beta\nu_2 & tr B_2 &= -(\lambda_1\alpha\mu_2 + \lambda_2\beta\nu_2). \end{aligned}$$

Setting the traces equal gives

$$\alpha(\mu_1-1)+\beta(\nu_1-1)=0, \qquad \alpha(\mu_2-1)+\beta(\nu_2-1)=0,$$
$$\lambda_1\alpha(\mu_1-1)+\lambda_2\beta(\nu_1-1)=0, \qquad \lambda_1\alpha(\mu_2-1)+\lambda_2\beta(\nu_2-1)=0.$$

Since $\lambda_1 \neq \lambda_2$, the only solution is $\mu_1=\mu_2=\nu_1=\nu_2=1$.

We have thus shown that $\langle A_i e_1, e_1\rangle = \langle A_i e_2, e_2\rangle = 1$ and $\langle A_i e_1, e_2\rangle = 0$ for $i=1,2$. It follows that $A_1 = A_2 = A$ and $B_1 = B_2 = B$. Therefore the point (A,B,C) under consideration is an extreme point. □

We have been unable to find a microstructure corresponding to any extreme point of type 3. However, the extreme points of types 1 and 2 are easily realized using sequential lamination. To achieve (4.12), we first layer pure phases 1 and 2 in volume fractions $\rho=\theta_1/(\theta_1+\theta_2)$ and $1-\rho=\theta_2/(\theta_1+\theta_2)$ using layers normal to k. Representing F as a block matrix as in (4.1), the layering formula (3.14) gives

$$F=\rho(1-\rho)\begin{pmatrix} k\otimes k & -k\otimes k & 0\\ -k\otimes k & k\otimes k & 0\\ 0 & 0 & 0\end{pmatrix}$$

for the result. Its volume fractions are of course $(\rho,1-\rho,0)$. Now we layer pure phase 3 with this, using volume fractions θ_3 and $1-\theta_3$ respectively, and still using layers normal to k. The result has volume fractions

$$\theta_3(0,0,1)+(1-\theta_3)(\rho,1-\rho,0)=(\theta_1,\theta_2,\theta_3).$$

Its geometric parameters are seen, after some calculation, to be in agreement with (4.12).

The construction leading to a type 2 extreme point is similar. To achieve (4.13) we first layer pure phases 2 and 3 in volume fractions $\rho=\theta_2/(\theta_2+\theta_3)$ and $(1-\rho)=\theta_3/(\theta_2+\theta_3)$ respectively, using layers normal to l. The result has volume fractions $(0,\rho,1-\rho)$ and geometric parameters

$$F=\rho(1-\rho)\begin{pmatrix} 0 & 0 & 0\\ 0 & l\otimes l & -l\otimes l\\ 0 & -l\otimes l & l\otimes l\end{pmatrix}.$$

Now we layer pure phase 1 with this, using volume fractions θ_1 and $1-\theta_1$ respectively, and using layers normal to k. The result has volume fractions

$$\theta_1(1,0,0)+(1-\theta_1)(0,\rho,1-\rho)=(\theta_1,\theta_2,\theta_3).$$

Its geometric parameters are seen, after some calculation, to be in agreement with (4.13). An analogous construction can be used to achieve (4.14).

To understand the nature of the type 3 extreme points better, it is interesting to consider those for which A, B, and C are simultaneously diagonal. This class will play a special role in Section 5. They have the form

$$\begin{pmatrix} A & B\\ B & C\end{pmatrix}=\begin{pmatrix} \begin{pmatrix} a_1 & 0\\ 0 & a_2\end{pmatrix} & \begin{pmatrix} b_1 & 0\\ 0 & b_2\end{pmatrix}\\ \begin{pmatrix} b_1 & 0\\ 0 & b_2\end{pmatrix} & \begin{pmatrix} b_1^2/a_1 & 0\\ 0 & b_2^2/a_2\end{pmatrix}\end{pmatrix} \tag{4.20}$$

with $a_1 > 0$, $a_2 > 0$, $b_1 \neq 0$, $b_2 \neq 0$, and $b_1/a_1 \neq b_2/a_2$. The trace relations require

$$\begin{aligned} a_1 + a_2 = \theta_1(1-\theta_1), \quad b_1 + b_2 = -\theta_1\theta_2, \\ b_1^2/a_1 + b_2^2/a_2 = \theta_2(1-\theta_2). \end{aligned} \tag{4.21}$$

A convenient way to classify the solutions of (4.21) is to seek them in the form $a_1 = \lambda b_1$, $a_2 = \mu b_2$, for some $\lambda \neq \mu \in \mathbf{R}$. Then (4.21) becomes

$$\begin{aligned} b_1 + b_2 = -\theta_1\theta_2, \quad \lambda b_1 + \mu b_2 = \theta_1(1-\theta_1), \\ \lambda^{-1}b_1 + \mu^{-1}b_2 = \theta_2(1-\theta_2). \end{aligned}$$

These are three independent equations in two unknowns b_1, b_2. They are consistent precisely if

$$(\lambda+1)(\mu+1)\theta_1\theta_2 + \theta_1\theta_3 + \lambda\mu\theta_2\theta_3 = 0, \tag{4.22}$$

and in that case they imply

$$\begin{aligned} a_1 &= \frac{\lambda}{\lambda-\mu}\left[\theta_1\theta_3 + (\mu+1)\theta_1\theta_2\right] \\ a_2 &= \frac{-\mu}{\lambda-\mu}\left[\theta_1\theta_3 + (\lambda+1)\theta_1\theta_2\right]. \end{aligned} \tag{4.23}$$

We want $a_1 \neq 0$ and $a_2 \neq 0$, so λ and μ should be non-zero. We also want $\lambda \neq \mu$. If

$$(\lambda+1)(\mu+1) \geq 0 \text{ and } \lambda\mu > 0 \tag{4.24}$$

then each term in (4.22) must vanish separately; it is easy to see that this cannot yield an extreme point of type 3. If however

$$(\lambda+1)(\mu+1) < 0 \quad \text{or} \quad \lambda\mu < 0 \tag{4.25}$$

then (4.22) has a one-parameter family of solutions $\theta \in V$. A tedious but straightforward calculation shows that the resulting values of a_1 and a_2, given by (4.23), are always positive. We have thus proved:

PROPOSITION 4.4. *Let λ and μ be given, with $\lambda \neq 0$, $\mu \neq 0$, and $\lambda \neq \mu$. If (4.24) holds then there is no type 3 extreme point of the form (4.20) with $a_1 = \lambda b_1$ and $a_2 = \mu b_2$. If on the other hand (4.25) holds, then there are type 3 extreme points of the form (4.20) with $a_1 = \lambda b_1$ and $a_2 = \mu b_2$. In fact there is one such extreme point in $\mathfrak{F}_\theta^U$ whenever $\theta \in V$ satisfies (4.22).*

We chose this section with an observation that will be useful in Section 5.

LEMMA 4.5. *The map $F_{ij\alpha\beta} \mapsto \Gamma_{ij}\delta_{\alpha\beta} - F_{ij\alpha\beta}$ takes type k extreme points to type k extreme points for each $k = 1,2,3$.*

Proof. We know from Corollary 3.4 that this map preserves $\mathcal{F}_\theta^U$. It is also affine and invertible, so it takes extreme points to extreme points. One verifies readily that $F_{11\alpha\beta}$ has rank 2 exactly if $\Gamma_{11}\delta_{\alpha\beta} - F_{11\alpha\beta}$ has rank 2, and similarly for $F_{22\alpha\beta}$. The desired conclusion follows easily from the character of the extreme points, c.f. (4.12) - (4.14). □

5. The three gradient problem. In this section we consider W of the form (1.3), with three wells in two space dimensions, under the further hypothesis that $a^i - a^j$ has rank 2 for each $i \neq j$. Our attention is focused on the question: is QW strictly positive for $\xi \notin \{a^i\}$? This question remains open, despite recent progress by Pedregal [23] and Šverák [27]. We do not solve it. Rather, we show a direct link between this problem and the attainability of our type 3 extreme points.

We begin with a digression, to set this discussion in its proper mathematical context. One reason for studying this "three gradient problem" is its analogy to the variational theory of phase transitions. In that setting it is an important task to identify all "macroscopically stress-free states." In other words, suppose that W has the form (1.1), and that each W_i has minimum value zero. Then one wants to know the exact set where $QW(\xi) = 0$. This is among the principal goals of [4,5]; there each W_i is assumed to be frame-in different, as is appropriate for geometrically nonlinear elasticity. The analogous question in a geometrically linear setting is addressed for example in [6] and [17]. Our attention here is on the corresponding question for "gradients," i.e. for energies of the form (1.3).

The "three gradient problem" is also of interest because it promises to shed light on the relationship between polyconvexity, rank-one convexity, and quasiconvexity. It is well-known that $PW \leq QW \leq RW$, where PW is the polyconvexification of W and RW is the rank-one convexification of W [9]. For a three-well energy with appropriately chosen "wells" $\{a^i\}$, it can happen that PW vanishes on a one-dimensional curve in the space of all matrices, while $RW(\xi) > 0$ for $\xi \notin \{a^i\}$ [23,27]. Knowing where QW vanishes could give some indication of whether $QW = PW$ or $QW = RW$ in this case.

Some of the relevant mathematical literature avoids discussing a relaxed energy, focussing instead on other concepts such as Young measures. To clarify the relation between such work and ours, we note some consequences of the assertion that $QW(\xi) = 0$. First, from the definition (1.1), if $QW(\xi) = 0$ then for every $\varepsilon, \delta > 0$ there exists $u : \Omega \to \mathbf{R}^m$ with affine boundary values $u|_{\partial\Omega} = \xi \cdot x$ such that

$$\text{meas}\{x : |\nabla u - a^i|^2 \geq \varepsilon \text{ for all } i\} \leq \delta. \tag{5.1}$$

Thus, loosely speaking, if $QW(\xi) = 0$ then a gradient field can have average value ξ and yet "approximately take only the values $a^i, 1 \leq i \leq N$." A second conclusion amounts to the same thing in different language: if $QW(\xi) = 0$ then there is a Young-measure limit of gradients with mean value ξ, whose support is contained in $\{a_i\}_{1\leq i\leq N}$. This is immediate from (1.2), by taking the Young-measure limit of a minimizing sequence. (See [3] and the references there for basic facts about Young measures, and [16] for more about the relation between quasiconvexity and Young measures.) The statement that $QW(\xi) = 0$ actually carries somewhat more information than (5.1). Among energies with super- or sub-quadratic growth

$$W(\nabla u) = \min_{1 \leq i \leq N} \{|\nabla u - a^i|^p\},$$

one might expect the set where $QW(\xi) = 0$ to depend on the value of p. Actually, it does not: motivated by a result of Šverák [26], Zhang [30] has shown, (under the hypothesis that the set where $W(\xi) = 0$ has compact support, which certainly applies in the case we are considering), that the set where $QW(\xi) = 0$ is independent of $1 \leq p < \infty$. In any case, we focus here on $p = 2$ because that is the case in which Fourier transform-based methods apply.

In any spatial dimension, and for any number of wells, we say that a^i and a^j are *compatible* if their difference has rank one. It is easy to see that if a^i and a^j are compatible for some $i \neq j$, then QW vanishes along the line segment joining a^i to a^j. The associated microstructure is a layered mixture of phases i and j with layer normal ν, where $a^i - a^j = w \otimes \nu$. It is natural to ask whether the converse holds.

Question 5.1: If $\{a^i\}_{1 \leq i \leq N}$ are pairwise incompatible, does it follow that $QW(\xi) > 0$ except when $\xi \in \{a^i\}$?

The answer is yes for $N = 2$ [17,20,24]. Surprisingly, it is no for $N = 4$ [6]. The case $N = 3$, which is the focus of our attention here, remains open in general.

In view of the discussion above concerning Young measures, etc., we think of the assertion that $QW(\xi) = 0$ for some $\xi \notin \{a^i\}$ as a statement that $\{a^i\}$ are *mutually compatible* (as gradients). Thus Question 5.1 asks whether a set of pairwise incompatible gradients is automatically mutually incompatible. See Section 8 of [17] for a related discussion in terms of microstructures.

Since the study of multiwell energies is motivated in part by the theory of coherent phase transitions, it seems appropriate to mention that the situation for strains is different from that for gradients. For geometrically linear elasticity, (1.3) should be replaced by

$$W(e) = \min_{1 \leq i \leq N} \{|e - a^i|^2\}$$

where $e(u) = (\nabla u + \nabla u^T)/2$ is a linear strain and $\{a^i\}$ are now symmetric $n \times n$ matrices. In this setting a^i and a^j are compatible if $a^i - a^j = v \otimes w + w \otimes v$ for some $v, w \in \mathbf{R}^n$. In space dimension 2, analogue of Question 5.1 has an affirmative answer for *any* number of wells. A similar result even holds in the geometrically nonlinear setting [6,10,21]. The situation in space dimension three is open, except for some special cases.

We return now to energies of the form (1.3). For the remainder of this section W is a three-well energy, involving incompatible, 2×2 matrices $\{a^i\}$. The following result gives an affirmative answer to Question 5.1 under certain conditions on $\{a^i\}$. The proof uses only the weak continuity of the determinant. Essentially the same result figures in the recent work of Pedregal and Šverák [23,27].

PROPOSITION 5.2. *Let a^1, a^2, and a^3 be pairwise incompatible 2×2 matrices. Assume that $\det(a^1 - a^2)$, $\det(a^2 - a^3)$, and $\det(a^1 - a^3)$ all have the same sign. Then $QW(\xi) > 0$ except when $\xi \in \{a^i\}$.*

Proof. We use Dacorogna's formula for the polyconvexification [8]:

$$PW(\xi) = \inf_{\nu \in \mathcal{A}_\xi} \int W(\lambda) d\nu(\lambda), \tag{5.2}$$

where ν ranges over the set $\mathcal{A}_\xi$ of probability measures on 2×2 matrices that satisfy the "minors relations":

$$\mathcal{A}_\xi = \{\nu : \int \lambda d\nu(\lambda) = \xi, \quad \int \det \lambda d\nu(\lambda) = \det \xi\}.$$

It is easy to see that the infimum in (5.2) is achieved, since W has quadratic growth at infinity.

Suppose that $QW(\xi) = 0$. It follows that $PW(\xi) = 0$, and therefore any extremal for (5.2) must be supported on the set where $W = 0$. We thus conclude the existence of

$$\nu = \sum_{i=1}^{3} \theta_i \delta_{a^i}, \quad \theta_i \geq 0, \quad \sum \theta_i = 1,$$

such that

$$\int \lambda d\nu(\lambda) = \sum \theta_i a^i = \xi$$

$$\int \det \lambda d\nu(\lambda) = \sum \theta_i \det(a^i) = \det \xi.$$

Combining these relations gives

$$\sum \theta_i \det(a^i) = \det(\sum \theta_i a^i). \tag{5.3}$$

Since we are in two dimensions, $\det(a)$ is a quadratic form and

$$\det(\sum \theta_i a^i) = \sum_i \theta_i^2 \det a^i + \sum_{i<j} \theta_i \theta_j \langle a^i, \text{ cof } a^j \rangle.$$

Using this in (5.3), we obtain after some manipulation that

$$\theta_1 \theta_2 \det(a^1 - a^2) + \theta_1 \theta_3 \det(a^1 - a^3) + \theta_2 \theta_3 \det(a^2 - a^3) = 0. \tag{5.4}$$

By hypothesis the terms $\det(a^i - a^j)$ are either all positive or all negative. So it follows that $\theta_1\theta_2 = \theta_1\theta_3 = \theta_2\theta_3 = 0$, whence $\xi = a^i$ for some i. □

What happens when $\{\det(a^i - a^j)\}$ are not all of the same sign? The answer is not known, however Pedregal and Šverák have proved the following [23,27]:

PROPOSITION 5.3. *Let a^1, a^2, and a^3 be incompatible 2×2 matrices. If $QW(\xi) = 0$ for some $\xi \notin \{a^i\}$, then the associated microstructure cannot be sequentially laminated. Equivalently, the rank-one convexification RW satisfies $RW(\xi) > 0$ whenever $\xi \notin \{a^i\}$.*

We turn now to the relationships between these issues and the calculus of geometric parameters. To keep matters as simple as possible, we shall consider only the case of ***diagonal*** matrices a^i. By a translation and a linear change of variables, there is no further loss of generality in taking $a^3 = 0$ and $a^1 = I$. So we shall henceforth concentrate on the case

$$a^1 = \begin{pmatrix} 1 & 0 \\ 0 & 1 \end{pmatrix}, \quad a^2 = \begin{pmatrix} -\lambda & 0 \\ 0 & -\mu \end{pmatrix}, \quad a^3 = \begin{pmatrix} 0 & 0 \\ 0 & 0 \end{pmatrix}. \tag{5.5}$$

The real numbers λ and μ should satisfy

$$(\lambda + 1)(\mu + 1) < 0 \quad \text{or } \lambda\mu < 0, \tag{5.6}$$

since otherwise Proposition 5.2 applies. (Our notation is chosen for later convenience in making contact with Proposition 4.4.)

THEOREM 5.4. *Let $\{a^i\}$ be as above. Then $QW(\xi) = 0$ with $\xi \notin \{a^i\}$ if and only if a certain associated type 3 extreme point of $\mathcal{F}_\theta^U$ is achievable. The specific extreme point is given by (5.16)-(5.17) below.*

Proof. Suppose that $QW(\xi) = 0$. Then obviously $CW(\xi) = 0$, so ξ lies in the convex hull of $\{a^i\}$:

$$\xi = \sum \theta_i a^i, \quad \theta_i \geq 0, \quad \sum \theta_i = 1. \tag{5.7}$$

Therefore ξ is necessarily diagonal. The "volume fractions" $\theta = (\theta_1, \theta_2, \theta_3)$ are uniquely determined by (5.7) since we have only three wells. The possible values of θ are restricted by (5.4), which becomes

$$\theta_1\theta_3 + \lambda\mu\theta_2\theta_3 + (\lambda + 1)(\mu + 1)\theta_1\theta_2 = 0 \tag{5.8}$$

in this case. In view of (5.6), the possible values of ξ lie on a one-dimensional curve. (An explicit representation of this curve is easy enough to obtain, but we omit it.)

We assert that in this situation there exists a tensor of geometric parameters $F_{ij\alpha\beta} \in \mathcal{F}_\theta$ satisfying

$$\sum (\Gamma_{ij}\delta_{\alpha\beta} - F_{ij\alpha\beta}) a^i_{\gamma\alpha} a^j_{\gamma\beta} = 0. \tag{5.9}$$

More specifically, we assert (5.9) whenever F represents the limiting behavior of a minimizing sequence in the definition of $QW(\xi)$. Indeed, for such F (2.10) - (2.11) yield

$$QW(\xi) = \sum \rho_i |\xi - a^i|^2 - \sum F_{ij\alpha\beta} a^i_{\gamma\alpha} a^j_{\gamma\beta}, \tag{5.10}$$

where $\rho = (\rho_1, \rho_2, \rho_3) \in V$ gives the volume fraction of the microstructure. Since $QW(\xi) = 0$, (5.10) and (3.11) yield

$$|\xi - \sum \rho_i a^i|^2 + \sum (\Gamma_{ij}\delta_{\alpha\beta} - F_{ij\alpha\beta}) a^i_{\gamma\alpha} a^j_{\gamma\beta} = 0. \tag{5.11}$$

The first term is a perfect square, and the second one is nonnegative as well, by (3.6). We therefore conclude that $\xi = \sum \rho_i a^i$, whence $\rho_i = \theta_i$. We furthermore conclude that F satisfies (5.9).

Next, we assert that any F satisfying (5.9) must be a type 3 extreme point of $\mathcal{F}^U_\theta$. Indeed, if (5.9) holds then

$$\inf_{F \in \mathcal{F}^U_\theta} \sum (\Gamma_{ij}\delta_{\alpha\beta} - F_{ij\alpha\beta}) a^i_{\gamma\alpha} a^j_{\gamma\beta} = 0, \tag{5.12}$$

using the fact that $\mathcal{F}_\theta \subset \mathcal{F}^U_\theta$ along with (3.6). The extremal value of any affine function on a compact, convex set is achieved at an extreme point; therefore F is either an extreme point of $\mathcal{F}^U_\theta$ or else a convex combination of extreme points all of which achieve 0 in (5.12). Now, there is no extreme point of type 1 or 2 that achieves 0 in (5.12). This can be proved by direct calculation using the formulae (4.12) - (4.14). Alternatively, it suffices to recall that these extreme points correspond to rank-two laminates, whereas it is impossible to achieve $QW(\xi) = 0$ by means of a rank-two lamination construction. We shall show presently that (5.12) vanishes (for given ξ) at a *unique* type 3 extreme point of $\mathcal{F}^U_\theta$. So F can satisfy (5.9) only by being equal to this type 3 extreme point.

Consider a type 3 extreme point $G \in \mathcal{F}^U_\theta$ which achieves 0 in (5.12). By Lemma 4.5,

$$H_{ij\alpha\beta} = \Gamma_{ij}\delta_{\alpha\beta} - G_{ij\alpha\beta} \tag{5.13}$$

is also an extreme point of type 3. Writing it as a block matrix

$$H = \begin{pmatrix} H_{11} & H_{12} \\ H_{12} & H_{22} \end{pmatrix} = \begin{pmatrix} A & B \\ B & C \end{pmatrix}$$

as in Section 4, we have $C = BA^{-1}B$, and AB^{-1} is not a multiple of the identity. The hypothesis that G achieves 0 in (5.12) can be written as

$$\langle Aa^1, a^1 \rangle + 2\langle Ba^1, a^2 \rangle + \langle Ca^2, a^2 \rangle = 0. \tag{5.14}$$

Because $C = BA^{-1}B$, (5.14) is a perfect square, i.e. (5.14) is equivalent to

$$|A^{1/2}a^1 + A^{-1/2}Ba^2|^2 = 0.$$

So $B^{-1}Aa^1 = -a^2$, i.e.

$$B^{-1}A = \begin{pmatrix} \lambda & 0 \\ 0 & \mu \end{pmatrix}. \tag{5.15}$$

Since A and B are symmetric, (5.15) implies that they are diagonal. To see this, we observe that $B^{-1}A = -a^2$ implies $A = -Ba^2$, so

$$A = A^T \Longrightarrow Ba^2 = a^2 B,$$

using the symmetry of a^2 as well as that of A and B. Thus B and a^2 commute, whence either $\lambda = \mu$ or else B is simultaneously diagonal with a^2. The case $\lambda = \mu$ does not arise, since $B^{-1}A$ cannot be a multiple of the identity at a type 3 extreme point. Thus B and $A = -Ba^2$ are both diagonal, and H has the form (4.20) with $a_1 = \lambda b_1$, $a_2 = \mu b_2$. By Proposition 4.4 there is exactly one such extreme point in $\mathfrak{F}_\theta^U$ whenever θ satisfies (5.8), and it is given by (4.23). An easy calculation shows that if H is given by (4.23), then $G_{ij\alpha\beta} = \Gamma_{ij}\delta_{\alpha\beta} - H_{ij\alpha\beta}$ is given by

$$G = \begin{pmatrix} \begin{pmatrix} c_1 & 0 \\ 0 & c_2 \end{pmatrix} & \begin{pmatrix} d_1 & 0 \\ 0 & d_2 \end{pmatrix} \\ \begin{pmatrix} d_1 & 0 \\ 0 & d_2 \end{pmatrix} & \begin{pmatrix} d_1^2/c_1 & 0 \\ 0 & d_2^2/c_2 \end{pmatrix} \end{pmatrix} \tag{5.16}$$

with

$$\begin{aligned} c_1 &= \frac{\mu}{\mu - \lambda}[\theta_1\theta_3 + (\lambda + 1)\theta_1\theta_2] \\ c_2 &= \frac{-\lambda}{\mu - \lambda}[\theta_1\theta_3 + (\mu + 1)\theta_1\theta_2] \\ d_1 &= \mu^{-1}c_1, \quad d_2 = \lambda^{-1}c_2. \end{aligned} \tag{5.17}$$

(Notice that (5.16) - (5.17) are simply (4.20) - (4.23) with the roles of λ and μ reversed.) Recapitulating, we have shown that if $QW(\xi) = 0$ then $\xi = \sum \theta_i a^i$ with $\theta \in V$ satisfying (5.8), and the associated microstructure achieves the tensor of geometric parameters determined by (5.16) - (5.17), which is an extreme point of type 3. The argument is entirely reversible: if $\theta \in V$ satisfies (5.8) and if the tensor of geometric parameters (5.16) - (5.17) is achievable by a microstructure, then retracing the argument leads to $QW(\xi) = 0$ with $\xi = \sum \theta_i a^i$. □

COROLLARY 5.5. *Type-3 extreme points with A, B, and C simultaneously diagonal cannot be achieved by laminated microstructures.*

Proof. Suppose that such a type 3 extreme point were achievable. Then by Theorem 5.4 we could obtain an example of a^1, a^2, and a^3 which are pairwise incompatible but mutually compatible. According to Proposition 5.3 the associated microstructure could not be sequentially laminated. □

Remark 5.6. The proof of Theorem 5.4 actually includes a new proof of Proposition 5.2, based on the "Geometric Parameters Lower Bound" instead of polyconvexification, when a^1, a^2, and a^3 are simultaneously diagonal. Indeed, the general case is easily reduced that of $\tilde{a}^3 = 0$, $\tilde{a}^1 = I$, $\tilde{a}^2 = (a^2 - a^3)(a^1 - a^3)^{-1}$. Condition (5.8) for $\tilde{a}^1, \tilde{a}^2, \tilde{a}^3$ is the same as (5.4) for a^1, a^2, and a^3.

REFERENCES

[1] ACERBI, E. AND FUSCO, N., *Semicontinuity problems in the calculus of variations*, Arch. Rat. Mech. and Anal., 86 (1984), pp. 125–145.

[2] AVELLANEDA, M., *Optimal bounds and microgeometrics for elastic two-phase composites*, SIAM J. Appl. Math. 47 (1987), pp. 1216–1228.

[3] BALL, J., *A version of the fundamental theorem for Young Measures*, in Partial Differential Equations and Continuum Models of Phase Transitions, M. Rascle et al. eds., Lecture Notes in Physics No. 344, Springer-Verlag (1989), pp. 207–215.

[4] BALL, J. AND JAMES, R., *Fine phase mixtures as minimizers of energy*, Arch. Rat. Mech. and Anal., 100 (1987), pp. 13–52.

[5] BALL, J. AND JAMES, R., *Proposed experimental tests of a theory of fine microstructure and the two-well problem*, preprint, 1991.

[6] BHATTACHARYA, K.; FIROOZYE, N.; JAMES, R.; AND KOHN, R., *Restriction on microstructure*, in preparation.

[7] DACOROGNA, B., *Quasiconvexity and relaxation of nonconvex problems in the calculus of variations*, J. Func. Anal., 46 (1982), pp. 102–118.

[8] DACOROGNA, B., *Remarques sur les notions de polyconvexité, quasiconvexité et convexité de rang 1*, J. Math. Pures Appl., 64 (1985), pp. 403–438.

[9] DACOROGNA, B., *Direct Methods in the Calculus of Variations*, Springer–Verlag, Berlin 1989.

[10] FIROOZYE, N., *Optimal Translations and Relaxations of Some Multiwell Energies*, Ph.D. Thesis, NYU, 1990.

[11] FIROOZYE, N., *On the relaxation of a class of isotropic, frame-indifferent elastic energies*, in preparation.

[12] FIROOZYE, N., *Optimal use of the translation method and relaxations of variational problems*, Comm. Pure. Appl. Math., to appear.

[13] GERARD, P., *Moyennisation et regularité 2-microlocale*, Ann. Scient. de l'Ecole Norm. Sup., to appear.

[14] JAMES, R. AND KINDERLEHRER, D., *Theory of diffusionless phase transitions*, in Partial Differential Equations and Continuum Models of Phase Transitions, M. Rascle et al. eds., Lecture Notes in Physics No. 344, Springer-Verlag (1989), pp. 51–84.

[15] KHACHATURYAN, A., *Theory of Structural Transformations in Solids*, John Wiley and Sons, New York, 1983.

[16] KINDERLEHRER, D. AND PEDREGAL, P., *Weak convergence of integrands and the Young measure representation*, I.M.A. Preprint #700, Sept 1990.

[17] KOHN, R., *The relaxation of a double well energy*, Continuum Mechanics and Thermodynamics, to appear.

[18] KOHN, R. AND STRANG, G., *Optimal design and relaxation of variational problems, I-III*, Comm. Pure Appl. Math., 39 (1986), pp. 113–137, 139–182, and 353–377.

[19] KOHN, R. AND VOGELIUS, M., *Relaxation of a variational method for impedance computed tomography*, Comm. Pure Appl. Math., 40 (1987), pp. 745–777.

[20] LURIE, K. AND CHERKAEV, A., *On a certain variational problem of phase equilibrium*, in Material Instabilities in Continuum Mechanics, J.M. Ball ed., Oxford Univ. Press (1988), pp. 257–268.

[21] MATOS, J., *Some Mathematical Methods of Mechanics*, Ph.D. Thesis, Univ. of Minn., 1990.

[22] MILTON, G. AND NESI, V., *Polycrystalline configurations that maximize electrical resistivity*, J. Mech. Phys. Solids, to appear.

[23] PEDREGAL, P., *Topics in Non-Linear Analysis*, Ph.D. Thesis, Univ. of Minn., 1989.

[24] PIPKIN, A., *Elastic materials with two preferred states*, Q.J. Mech. Appl. Math., to appear.

[25] ROITBURD, A., *Martensitic transformation as a typical phase transformation in solids*, in Solid State Physics 33, Academic Press (1978), pp. 317–390.

[26] ŠVERÁK, V., *Quasiconvex functions with subquadratic growth at infinity*, preprint, 1990.

[27] ŠVERÁK, V., to appear.

[28] TARTAR, L., *Estimations fines des coefficients homogégéisés*, in Ennio de Georgi's Colloquium, P. Krée ed., Pitman Research Notes in Mathematics (1985), pp. 168–187.

[29] TARTAR, L., *H-measures, a new approach for studying homogenization, oscillations, and concentration effects in partial differential equations*, Proç. Roy. Soc., Edinburgh, 115A (1990), pp. 193–230.

[30] ZHANG, K., *Construction of quasiconvex functions with linear growth at infinity*, preprint, 1990.

VARIATIONAL PROBLEMS FOR CRYSTALS WITH DEFECTS

IRENE FONSECA* AND GARETH PARRY**

Abstract. Equilibrium configurations for crystals with defects are analyzed. Using the theory of compensated compactness the state functions are characterized in the case where minimizing sequences develop oscillations. A new class of variational problems involving bulk and surface energy terms is studied.

Key words. Crystals with defects, neutral deformations.

AMS(MOS) subject classifications. 35D, 73G05, 49A50.

1. Introduction. In this work we will be dealing with crystals exhibiting so few defects that a lattice is distinguishable at the microscopic scale. A (perfect) *crystal* is a discrete array of identical atoms arranged in a periodic way and filling the space $\mathbb{R}^3$, precisely there exist lattice vectors a_1, a_2, a_3 with corresponding lattice matrix L such that $Le_i = a_i$ for $i = 1, 2, 3$[1] and the position vector of any atom is given by

$$x = m_1 a_1 + m_2 a_2 + m_3 a_3$$

with $m_1, m_2, m_3 \in \mathbb{Z}$. Clearly L is not unique and L' is another lattice matrix if and only if

$$L' = LH \tag{1.1}$$

for $H \in SL_3(\mathbb{Z}) := \{H \in M^{3\times 3} | \det H = \pm 1, H_{ij} \in \mathbb{Z}, i, j = 1, 2, 3\}$ (see ERICKSEN [7], FONSECA [8], KINDERLEHRER [15]). A *crystal with defects* is an array of atoms that cannot be mapped globally onto a space lattice, such operation being possible only locally and perhaps with the exclusion of some atoms. In this case the lattice vectors are actually averages over microscopic regions.

To make the bridge between microscopic and macroscopic behaviors of the material, DAVINI [3], and in subsequent work DAVINI & PARRY [4], [5], introduced a kinematical model for crystals with defects where deformations that leave a certain class of elastic invariants unchanged preserve the defectiveness. Within this theory, we take the viewpoint that equilibria correspond to minimizers of an energy functional

$$I(L) := \int_\Omega W(L(x))dx$$

*Department of Mathematics, Carnegie Mellon University, Pittsburgh, PA 15213. The research of the first author was partially supported by the National Science Foundation under Grant No. DMS–8803315. This collaboration took place during a visit of I. Fonseca to the University of Bath (U.K.) in May-July 1990, supported by a visiting fellowship of the Science and Engineering Research Council of the U.K.

**University of Bath, School of Mathematical Sciences, Claverton Down, Bath, Avon, BA2 7AY, U.K.

[1] In what follows, $\{e_1, e_2, e_3\}$ is the canonical basis of $\mathbb{R}^3$.

where L is a lattice matrix corresponding to a defect-preserving deformation of a perfect crystal with reference configuration Ω, and W represents the bulk energy density. Due to (1.1) and to frame indifference, W must satisfy

$$(1.2)\qquad W(RLH) = W(L) \text{ for all rotation } R \in O^+(3), L \in M_+^{3\times 3}, H \in SL_3(\mathbb{Z}).$$

Since W is not quasiconvex (see ERICKSEN [7], FONSECA [8], KINDERLEHRER [15]), the energy $I(\cdot)$ is not sequentially weakly lower semicontinuous and so, in general, we do not expect existence of minimizers. Following the work by CHIPOT & KINDERLEHRER [2] corresponding to elastic changes of state, we are interested in studying the behavior of minimizing sequences and their state functions rather than the macroscopic weak limit. Assuming that solutions may be measure-valued and using the parametrized probability measures of YOUNG [22] and the theory of compensated compactness of MURAT & TARTAR (see TARTAR [16]), we are able to calculate the energy and stresses of the deformed body when the class of variations includes non-elastic changes.

In Section 2 we give a brief description of the model for crystals with defects proposed by DAVINI [3] and DAVINI & PARRY [4], [5]. Non-elastic defect-preserving deformations of a perfect crystal are called *neutral* and we show that, in the case of a perfect cubic crystal, the corresponding lattice matrix L can be written as

$$(1.3)\qquad L(u(x)) = \nabla u(x)\{\nabla v(x)\}^{-1},$$

where $u : \Omega \to \mathbb{R}^3$ is the elastic deformation and v represents the slip or plastic deformation, with $\det \nabla v = 1$ a.e. in Ω (see FONSECA & PARRY [12]). Using the div-curl lemma (see TARTAR [16]) we prove that the class of neutral deformations is closed with respect to the weak $*$ convergence in $W^{1,\infty}$ (see Proposition 2.7) and in Proposition 2.9 we show that the minors of L_n are weakly $*$ continuous.

Due to (1.3) and assuming Dirichlet boundary conditions, we rewrite the energy $I(.)$ as the functional

$$E(u,v) := \int_\Omega W(\nabla u(x)\{\nabla v(x)\}^{-1})dx$$

where (u, v) belong to the set of admissible pairs

$$\mathcal{A}(u_0) := \{(u,v) \in W^{1,\infty}(\Omega,\mathbb{R}^3) |\det \nabla u > 0 \text{ a.e. in } \Omega, u = u_0 \text{ on } \partial\Omega, \det \nabla v = 1 \text{ a.e. in } \Omega\}$$

which, when v is the identity map, includes the elastic deformations.

In Section 3 we prove that the relaxation of the energy $E(.,.)$ coincides with

$$\inf\left\{\int_\Omega g^{**}(\det \nabla u(x))dx | u \in W^{1,\infty}(\Omega,\mathbb{R}^3), \det \nabla u > 0 \text{ a.e. in } \Omega, u = u_0 \text{ on } \partial\Omega\right\}$$

where g^{**} is the convex minorant of the subenergy function of ERICKSEN and FLORY,

$$g(t) := \inf\{W(F) | \det F = t\}.$$

In Theorem 3.1 we show that the average weak limit of the Cauchy stress tensor corresponding to a minimizing sequence must be isotropic. ERICKSEN [6] had remarked that perfect elastic crystals cannot support shear stresses (see FONSECA [8]) and later CHIPOT & KINDERLEHRER [2] showed that the average Cauchy stress for an elastic crystal is also a multiple of the identity. Moreover, DAVINI & PARRY [4] proved that, if we allow neutrally related states to compete, then some of the stress averages vanish independently of the material symmetry hypotheses on W.

In Section 4 we modify $E(u, v)$ by introducing a perturbation which depends on the slip v and we investigate whether or not this penalization will affect the isotropy of the Cauchy stress. The total energy becomes

$$\int_{\Omega} W(\nabla u(x)\{\nabla v(x)\}^{-1})dx + \int_{\partial v(\Omega)} \Gamma(\nu)dS \tag{1.4}$$

where Γ denotes the surface tension and ν is the normal to $\partial v(\Omega)$. In Theorem 5.2 we prove that the infimum of (1.4) is equal to the sum of the infima, precisely

$$\int_{\Omega} \varphi^{**}(\det \nabla u_0(x))dx + \int_{\partial C} \Gamma(\nu)dS$$

where C is the Wulff shape for Γ (see HERRING [14], WULFF [21]). Thus, there is a decoupling of the elastic and plastic parts of the weak limit of the appropriate minimizing sequence and once more, the weak limit of the corresponding Cauchy stress tensor is isotropic. Even if the results do not quite fit with the physics as yet, this analysis seems to represent a new direction for the calculus of variations which is motivated by mainstream problems in the mechanics of solids.

2. Crystals with defects and neutral deformations. We give a brief description of a mathematical model for crystals with defects proposed by DAVINI [3] and DAVINI & PARRY [4], [5]. For more details see FONSECA & PARRY [12].

In the sequel, Ω is a bounded, open, strongly Lipschitz domain in $\mathbb{R}^3$ and it represents the macroscopic placement of a non-defective cubic crystal. $M^{3\times 3}$ denotes the space of real 3×3 matrices and $M_+^{3\times 3} := \{F \in M^{3\times 3} | \det F > 0\}$. Consider a change of state from the reference $\Sigma_0 := \{\Omega, \mathbf{1}\}$ to $\Sigma := \{u(\Omega), L\}$, where the macroscopic deformation $u : \Omega \to u(\Omega)$ is a Lipschitz function with $\det \nabla u > 0$ a.e. in Ω and $L(u(x))$ represents average values, over microscopic regions, of the lattice vectors around $u(x)$ defining the lattice cell at the atomic scale. In this theory the evolution of defects is supposed to account for the discrepancy between the macroscopic deformation and the lattice vectors as they come from averaging at different scales. The main feature of the work by DAVINI [3] and DAVINI & PARRY [4], [5] is the introduction of a class of changes of state which strictly includes the elastic ones and which, at least intuitively, leave the defectiveness unchanged: they correspond to elastic deformations followed by rearrangements at the microscopic level.

DEFINITION 2.1. We say that Σ_0 and Σ are *elastically related* if the Cauchy–Born hypothesis is satisfied, namely $L(u(x)) = \nabla u(x)$ a.e. $x \in \Omega$.

Since elastically related states preserve the defectiveness (see TAYLOR [17]), it is natural to search for those integrals of the type

$$\int_c f(\Delta).dx, \int_S f(\Delta).n(x)d\sigma(x), \qquad \int_V f(\Delta)dx$$

where $\Delta := \{L, \nabla L\}$ represents a *local state of the crystal*, which will remain invariant under elastic deformations. Here c is the boundary of a surface Π and S is the boundary of a volume region V. Naturally, the densities corresponding to these integrals will produce a list of defect measures.

We introduce some notation. The *lattice vectors* are given by

$$1_i(x) := L(x)e_i, i = 1,2,3$$

and the *dual lattice vectors* are defined by

$$d_i(x) := D(x)e_i, i = 1,2,3$$

where

$$D := L^{-T}.$$

Clearly

$$1_i(x).d_j(x) = \delta_{ij} \text{ and } 1_i(x) = \frac{1}{2}\varepsilon_{ijk}\frac{d_j \wedge d_k}{\det D} \text{ with } i,j,k \in \{1,2,3\}.$$

Also for $i, j \in \{1,2,3\}$ we define the following densities, where $y = u(x)$:

$$\begin{aligned}
b_i &:= \operatorname*{curl}_y d_i && (\textit{Burger's vectors})\\
\sigma_{ij} &:= b_i.d_j && (\text{components of } \textit{Bilby's dislocation tensor})\\
n &:= 1/\det L = \det D && (\text{the number of cells per u nit volume})\\
m &:= \frac{1}{n \det \nabla u(x)} && (\text{atomic mass of an average cell})\\
g_i &:= \nabla_y m.1_i &&\\
\delta_i &:= \nabla_y m \wedge d_i &&
\end{aligned}$$

THEOREM 2.2. ([3])
1. *(Invariants associated to line defects) A line integral*

$$\int_c f(\Delta).dx \tag{2.1}$$

is elastic invariant if and only if there exists a function $h : \mathbf{R} \to \mathbf{R}$ *such that*

$$f = h(m)d;$$

2. *(Invariants associated to point defects) A surface integral*

$$\int_S f(\Delta).n(x)dH_2(x) \tag{2.2}$$

where H_2 *is the 2-dimensional Hausdorff measure and* $n(x)$ *is the normal to the surface* S *at the point* x*, is elastic invariant if and only if there exists a function* $h : \mathbf{R} \to \mathbf{R}$ *such that*

$$f = h(m)d_j \wedge d_k;$$

3. *A volume integral*

$$\int_V f(\Delta)dx \tag{2.3}$$

is elastic invariant if and only if there exists a function $h : \mathbf{R}^3 \to \mathbf{R}$ *such that*

$$f = h(\sigma/n, m, g)n.$$

Further, it can be shown that if $\varphi \in \left\{\frac{\sigma}{n}, g, m\right\}$ *then the integrals*

$$\int_V (\nabla\varphi.l_i)ndx, \quad \int_\tau (\nabla\varphi.l_i)d_j.dx \;\; i,j = 1,2,3 \tag{2.4}$$

are elastic invariants.

It turns out that the class of states Σ for which the elastic invariants remain unchanged is strictly larger than the class of elastically related states (see Examples 2.5).

DEFINITION 2.3. The states Σ and Σ^* are said to be *neutrally related* if they are not elastically related although the integrals (2.1)–(2.4) remain invariant.

We obtain the following characterization for neutral states. For its proof we refer the reader to FONSECA & PARRY [12].

PROPOSITION 2.4. Σ_0 *and* $\Sigma = \{u(\Omega), L\}$ *are neutrally related if and only if there exists a Lipschitz mapping* $v : \Omega \to \mathbf{R}^3$ *such that for almost all* $x \in \Omega$

1. $L(u(x)) = \nabla u(x)\{\nabla v(x)\}^{-1}$;
2. $\det \nabla v(x) = 1$.

We give some examples of neutrally related states.

Examples 2.5.

(1) Set $\Sigma = \{u(\Omega), L\}$ where $u \in W^{1,\infty}(\Omega, \mathbf{R}^3)$ is invertible, $\det \nabla u(x) > 0$ a.e. in Ω,

$$L(u(x)) = \nabla u(x)(1 + a \otimes b)$$

where $a, b \in \mathbf{R}^3$ are such that $a.b = 0$. By Proposition 2.4 it follows that Σ_0 and Σ are neutrally related.

(2) Let $\Sigma = \{u(\Omega), \mathbf{1}\}$, where $u(x) = x + x_3 e_2 = (\mathbf{1} + e_2 \otimes e_3)x$ and $v = u$. Σ is a special type of *rearrangement* of Σ_0, designated by *slip* in the classic phenomenological plasticity theories.

It can be shown that if Σ_0 and Σ are neutrally related then they are locally elastically related, i.e. for all $x_0 \in \Omega$ there exist neighborhoods U_1 and U_2 of x_0 in Ω and there exists a Lipschitz function $g : U_1 \to U_2$ such that $g(x_0) = x_0$ and

$$L((u \circ g)(x)) = \nabla(u \circ g)(x) \text{ for almost all } x \in U_1.$$

Hence, it is natural to assume that neutrally related states preserve the defectiveness.

Next, we want to study equilibria of crystals within a variational framework where neutrally related states are admissible. As it is well known, the bulk energy density for solid crystals is non quasiconvex and so, in general, the energy functional is not lower semicontinuous as minimizing sequences may develop oscillations. In particular, the macroscopic limit is not necessarily a minimizer of $E(.,.)$ and, as it turns out, the sequence itself stores more information on limiting macroscopic state functions of the crystal than the macroscopic configuration itself. This information is given partially by the corresponding Young measure (see YOUNG [22] and TARTAR [16]) as it shows the work by CHIPOT & KINDERLEHRER [2] for elastic crystals. However, before we start the analysis of the Young measures associated to minimizing sequences, we need to make sure that these sequences are "stable" under weak convergence, even if oscillations may occur. We will prove this result using MURAT & TARTAR's div-curl lemma of the theory of compensated compactness (see TARTAR [16]).

DIV-CURL LEMMA 2.6. *Let $\Omega \subset \mathbf{R}^N$ be an open, bounded, strongly Lipschitz domain, and let $u_n, v_n \in L^\infty(\Omega; \mathbf{R}^N)$ be such that $u_n \to u_\infty$ and $v_n \to v_\infty$ in L^∞ weak $*$. If, in addition, $\{\text{div } u_n\}$ and $\{\text{curl } v_n\}$ are bounded sequences in $H^{-1}_{\text{loc}}(\Omega)$ then $u_n.v_n \to u_\infty.v_\infty$ weakly $*$.*

PROPOSITION 2.7. *Let $\Sigma_n = \{u_n(\Omega), L_n\}$ be a sequence of states neutrally related to Σ_0 and let $v_n : \Omega \to \mathbf{R}^3$ be such that $L_n(u(x)) = \nabla u_n(x)\{\nabla v_n(x)\}^{-1}$, $\det \nabla v_n(x) = 1$ for almost all $x \in \Omega$. If $u_n \to u_\infty$ and $v_n \to v_\infty$ in $W^{1,\infty}$ weak $*$ then*

1. *$\det \nabla v_\infty(x) = 1$ a.e. in Ω;*
2. *$L_n(x) \to L_\infty(x) := \nabla u_\infty(x)\{\nabla v_\infty(x)\}^{-1}$ in L^∞ weak $*$.*

Proof. (1) follows immediately from the fact that $F \to \det F$ is a null lagrangian. Also, as

$$\{\nabla v_n(x)\}^{-1} = \{\text{adj } \nabla v_n(x)\}^T$$

with the rows of $\nabla u_n(x)$ curl free and the rows of adj $\nabla v_n(x)$ divergence free[2], we obtain (2) from the div-curl lemma.

As we mentioned before, we are interested in the characterization of the Young measure associated to a bounded minimizing sequence of lattice matrices. We start by recalling the notion of parametrized probability measures.

PROPOSITION 2.8. *If $\{u_\eta\}$ is a bounded sequence in $L^\infty(\Omega, \mathbb{R}^p)$ then there exists a subsequence $\{u_\varepsilon\}$ and a family of probability measures $\{\mu_x\}_{x\in\Omega}$ (Young measure) such that if $f \in C(\mathbb{R}^p)$ then $\{f(u_\varepsilon)\}$ converges in L^∞ weak $*$ to the average function*

$$\bar{f}(x) := \langle \mu_x, f\rangle = \int_{\mathbb{R}^p} f(y) d\mu_x(y).$$

As in Proposition 2.7, consider a sequence of states $\Sigma_n = \{u_n(\Omega), L_n\}$ neutrally related to Σ_0 and let $v_n : \Omega \to \mathbb{R}^3$ be such that $L_n(u(x)) = \nabla u_n(x)\{\nabla v_n(x)\}^{-1}$, $\det \nabla v_n(x) = 1$ for almost all $x \in \Omega$. Let $u_n \to u_\infty$, $v_n \to v_\infty$ in L^∞ weak $*$, $L_n(x) \to L_\infty(x) := \nabla u_\infty(x)\{\nabla v_\infty(x)\}^{-1}$ in L^∞ weak $*$ and let $\{\mu_x\}_{x\in\Omega}$ be the Young's measure corresponding to $\{L_n\}$. If the change is elastic then $L_n = \nabla u_n$, $L_\infty = \nabla u_\infty$, and as $M \to \det(M)$ and $M \to \text{adj } M$ are null lagrangians it follows that

$$\det\left(\int_{M^{3\times 3}} M d\mu_x(M)\right) = \int_{M^{3\times 3}} \det M d\mu_x(M) \tag{2.5}$$

and

$$\text{adj}\left(\int_{M^{3\times 3}} M d\mu_x(M)\right) = \int_{M^{3\times 3}} \text{adj } M d\mu_x(M). \tag{2.6}$$

The analysis of CHIPOT & KINDERLEHRER [2] relies heavily on (2.5) and (2.6). We show that this properties still hold for neutrally related states.

PROPOSITION 2.9. *For almost all $x \in \Omega$ we have*

1. $\det \nabla u_\infty(x) = \det L_\infty(x) = \det\left(\int_{M^{3\times 3}} M d\mu_x(M)\right)$

$$= \int_{M^{3\times 3}} \det M d\mu_x(M);$$

2. $\text{adj } L_\infty(x) = \text{adj}\left(\int_{M^{3\times 3}} M d\mu_x(M)\right) = \int_{M^{3\times 3}} \text{adj } M d\mu_x(M).$

Proof. By Propositions 2.7 and 2.8 we have

$$L_\infty(x) = \int_{M^{3\times 3}} M d\mu_x(M)$$

[2] adj A is the matrix of cofactors of A. In particular, if A is invertible then $A^{-1} = \frac{(\text{adj } A)^T}{\det A}$. Also, the inner product between matrices is defined by $A.B := tr(A^T B)$.

and

$$\det\left(\int_{M^{3\times 3}} M d\mu_x(M)\right) = \det L_\infty(x)$$

$$= \det(\nabla u_\infty(x)\{\nabla v_\infty(x)\}^{-1})$$
$$= \det \nabla u_\infty(x)$$
$$= w.* \text{ limit } \det \nabla u_n(x)$$
$$= w.* \text{ limit } \det L_n(x)$$
$$= \int_{M^{3\times 3}} \det M d\mu_x(M)$$

and, in a similar way,

$$\text{adj}\left(\int_{M^{3\times 3}} M d\mu_x(M)\right) = \text{adj } L_\infty(x)$$

$$= \text{adj } \nabla u_\infty(x)\{\nabla v_\infty(x)\}^T$$

with the rows of adj ∇u_∞ divergence free and the rows of $\nabla\varphi_\infty(x)$ curl free. By the div-curl lemma we conclude that

$$\text{adj}\left(\int_{M^{3\times 3}} M d\mu_x(M)\right) = w.* \text{ limit adj } \nabla u_n(x)\{\nabla v_n(x)\}^T$$

$$= w.* \text{ limit adj } L_n(x)$$
$$= \int_{M^{3\times 3}} \text{adj } M d\mu_x(M).$$

3. Relaxation of the bulk energy. Here we take the viewpoint that crystal equilibria correspond to extremals of some energy functional. Taking into account Proposition 2.4, we assume that the bulk energy associated to a neutral state $\Sigma = \{u(\Omega), L\}$ is given by

$$E(u,v) = \int_\Omega W(\nabla u(x)\{\nabla v(x)\}^{-1})dx,$$

where $L(u(x)) = \nabla u(x)\{\nabla v(x)\}^{-1}$, $\det \nabla v(x) = 1$ a.e. in Ω, and W is the stored energy density satisfying (1.2). Also, as it is usual, in order to make it energetically impossible to compress part of the body to zero volume or to change its orientation we suppose that

$$W(L) \to 0^+ \text{ as } \det L \to 0^+. \tag{3.1}$$

Considering Dirichlet boundary conditions, Σ is an admissible change if $(u,v) \in \mathcal{A}(u_0)$ with

$$\mathcal{A}(u_0) := \{(u,v) \in W^{1,\infty}(\Omega, \mathbb{R}^3) \mid \det \nabla u > 0 \text{ a.e. in } \Omega, u = u_0 \text{ on } \partial\Omega, \det \nabla v = 1 \text{ a.e. in } \Omega\}.$$

This class includes the elastic deformations in the case where v is the identity map. Here $u_0 \in C^1(\overline{\Omega}, \mathbb{R}^3)$ is one-to-one in Ω, $\det \nabla u_0 > 0$ in Ω and

$$\inf\left\{\int_\Omega g^{**}(\det\nabla u(x))dx | u\in W^{1,\infty}(\Omega,\mathbb{R}^3), \det\nabla u>0 \text{ a.e. in } \Omega, u=u_0 \text{ on } \partial\Omega\right\}$$

$$=\inf\left\{\int_\Omega g^{**}(\det\nabla u(x))dx | u\in C^1(\overline{\Omega},\mathbb{R}^3), \det\nabla u>0 \text{ a.e. in } \Omega, u=u_0 \text{ on } \partial\Omega\right\}^3.$$

In particular, these conditions imply that if $(u,v)\in\mathcal{A}(u_0)$ then u is invertible a.e. (see BALL [1]). Using Lagrange multipliers, DAVINI & PARRY [4] showed that, independently of the boundary conditions and of the symmetry invariance, at a smooth local minimizer we must have

$$\int_\Omega L^T(x)S(L(x))dx=\alpha\mathbf{1}$$

where the first Piola-Kirchhoff stress tensor S and the Cauchy stress tensor T are given, respectively, by

$$S(F):=\frac{\partial W}{\partial F}(F) \text{ and } T(F)=\frac{1}{\det F}S(F)F^T \text{ for all } F\in M_+^{3\times3}.$$

Hence, they concluded that there is a weakness in the crystal associated to the presence of slips and rearrangements, as the crystal cannot sustain certain nonzero average stresses. Also, using the material symmetry invariance (1.2), ERICKSEN [6] proved that for elastic crystals at equilibrium the Cauchy stress reduces to a pressure,

$$T=-p\mathbf{1} \tag{3.2}$$

and later CHIPOT & KINDERLEHRER [2] recovered (3.2) still for elastic changes and when oscillations develop. Precisely, they showed that if $\{\mu_x\}_{x\in\Omega}$ is the Young's measure corresponding to a minimizing sequence $L_n=\nabla u_n$, where $u_n\to u_\infty$ in $W^{1,\infty}$ weak $*$, then the average Cauchy stress is still a pressure,

$$\begin{aligned}\overline{T}(x)&=\int_{M^{3\times3}} T(M)d\mu_x(M)\\ &=(g^{**})'(\det\nabla u_\infty(x))\mathbf{1} \text{ a.e. in } \Omega\end{aligned} \tag{3.3}$$

where g is the *subenergy function* introduced by ERICKSEN and FLORY,

$$g(t):=\inf\{W(F)|\det F=t\}.$$

Here we will show that (3.3) still holds even when neutral changes of state are allowed to compete. Let $\Sigma_n=\{u_n(\Omega),L_n\}$ be a minimizing sequence for $I(.)$ with $(u_n,v_n)\in\mathcal{A}(u_0)$ and

$$L_n(u_n(x))=\nabla u_n(x)\{\nabla v_n(x)\}^{-1}.$$

Suppose further that $u_n\to u_\infty$ and $v_n\to v_\infty$ in $W^{1,\infty}$ weak $*$, which, by Proposition 2.7 imply that $L_n\to L_\infty=\nabla u_\infty\{\nabla v_\infty\}^{-1}$ in L^∞ weak $*$. Let $\{\mu_x\}_{x\in\Omega}$ be the Young's measure associated to $\{L_n\}$.

---

[3] As an example, if $\det\nabla u_0=$ const. a.e. in Ω then using Jensen's inequality it is easy to check that this hypothesis is satisfied.

THEOREM 3.1.

$$
\begin{aligned}
\overline{T}(x) &= \int\limits_{M^{3\times 3}} T(M)d\mu_x(L) \\
&= \frac{1}{\det L_\infty(x)}\overline{S}(x)L_\infty^T \\
&= (g^{**})'(\det \nabla u_\infty(x))\mathbf{1} \text{ for almost all } x \in \Omega.
\end{aligned}
$$

We start by obtaining the relaxation of the bulk energy functional.

THEOREM 3.2. *Let* $A \in M_+^{3\times 3}$ *and let* $\mathcal{A}(A) := \{(\xi, v) \in W^{1,\infty}(\Omega, \mathbf{R}^3) \times W^{1,\infty}(\Omega, \mathbf{R}^3) | \xi(x) = Ax \text{ on } \partial\Omega, \det \nabla v(x) = 1 \text{ a.e. in } \Omega\}$. *Then*

$$
\inf\left\{\int\limits_\Omega W(\nabla\xi(x)\{\nabla v(x)\}^{-1})dx | (\xi, v) \in \mathcal{A}(A)\right\} = \operatorname{meas}(\Omega)g^{**}(\det A).
$$

This result was proven by CHIPOT & KINDERLEHRER [2] and FONSECA [9] in the case where only elastic changes are admissible:

$$
\inf\left\{\int\limits_\Omega W(\nabla\xi(x))dx | \xi \in Ax + W_0^{1,\infty}(\Omega, \mathbf{R}^3)\right\} = \operatorname{meas}(\Omega)g^{**}(\det A). \tag{3.4}
$$

Proof of Theorem 3.2. Clearly, by (3.4)

$$
\begin{aligned}
&\inf\left\{\int\limits_\Omega W(\nabla\xi(x)\{\nabla v(x)\}^{-1})dx | (\xi, v) \in \mathcal{A}(A)\right\} \leq \\
&= \inf\left\{\int\limits_\Omega W(\nabla\xi(x))dx | \xi \in Ax + W_0^{1,\infty}(\Omega, \mathbf{R}^3)\right\} = \operatorname{meas}(\Omega)g^{**}(\det A)
\end{aligned}
$$

and since $F \to \det F$ is a null lagrangian, by Jensen's inequality and as $W(F) \geq g^{**}(\det F)$ we have

$$
\begin{aligned}
\int\limits_\Omega W(\nabla\xi(x)\{\nabla v(x)\}^{-1})dx &\geq \int\limits_\Omega g^{**}(\det \nabla\xi(x) \det \nabla v(x)^{-1})dx \\
&= \int\limits_\Omega g^{**}(\det \nabla\xi(x))dx \\
&\geq \operatorname{meas}(\Omega)g^{**}\left(\frac{1}{\operatorname{meas}(\Omega)}\int\limits_\Omega \det \nabla\xi(x)dx\right) \\
&= \operatorname{meas}(\Omega)g^{**}(\det A).
\end{aligned}
$$

From Theorem 3.2 and using the same argument as in CHIPOT & KINDERLEHRER [2] we obtain the following generalization to the case of inhomogeneous boundary conditions.

PROPOSITION 3.3.

$$\inf\left\{\int_\Omega W(\nabla\xi(x)\{\nabla v(x)\}^{-1})dx|(\xi,v)\in\mathcal{A}(u_0)\right\}=$$

$$\inf\left\{\int_\Omega W(\nabla\xi(x))dx|\xi\in Ax+W_0^{1,\infty}(\Omega,\mathbf{R}^3)\right\}=\text{meas}\,(\Omega)g^{**}(\det A).$$

PROPOSITION 3.4.

1. $\int_\Omega \overline{W}(x)dx=\int_\Omega\left(\int_{M^{3\times3}} W(M)d\mu_x(M)\right)dx$

$$\leq\inf\left\{\int_\Omega W(\nabla\xi(x)\{\nabla v(x)\}^{-1})dx|(\xi,v)\in\mathcal{A}(u_0)\right\};$$

2. supp $\mu_x\subset M_+^{3\times3}$ *and* $\det\nabla u_\infty>0$ a.e. *in* Ω.

Proof. Part (1) is proven exactly as in CHIPOT & KINDERLEHRER [2] and (3.1) and (1) imply that supp $\mu_x\subset M_+^{3\times3}$. Finally, by Theorem 3.2

$$\begin{aligned}\nabla u_\infty(x)\{\nabla v_\infty(x)\}^{-1}&=L_\infty(x)\\&=\int_{M^{3\times3}} M\,d\mu_x\end{aligned}$$

which, together with Proposition 2.9 (1), yields

$$\det\nabla u_\infty(x)=\int_{M^{3\times3}}\det M\,d\mu_x>0\text{ a.e. in }\Omega.$$

COROLLARY 3.5. *Under the hypotheses of Proposition 3.3*

$$\inf\left\{\int_\Omega W(\nabla\xi(x)\{\nabla v(x)\}^{-1})dx|(\xi,v)\in\mathcal{A}(u_0)\right\}=\int_\Omega g^{**}(\det\nabla u_\infty(x))dx.$$

Proof. Using the same argument as in CHIPOT & KINDERLEHRER [2], by Proposition 3.4 we have $\det\nabla u_\infty>0$ a.e. in Ω and so, since g is convex and $\det\nabla u_n\to\det\nabla u_\infty$ weakly $*$, by Proposition 3.3 we conclude that

$$\begin{aligned}\inf\left\{\int_\Omega W(\nabla\xi(x)\{\nabla v(x)\}^{-1})dx|(\xi,v)\in\mathcal{A}(u_0)\right\}&\leq\int_\Omega g^{**}(\det\nabla u_\infty(x))dx\\&\leq\liminf\int_\Omega g^{**}(\det\nabla u_n(x))dx\\&\leq\liminf\int_\Omega W(L_n(x))dx\\&=\inf\left\{\int_\Omega W(\nabla\xi(x)\{\nabla v(x)\}^{-1})dx|(\xi,v)\in\mathcal{A}(u_0)\right\}.\end{aligned}$$

PROPOSITION 3.6.
1. $\overline{W}(x) = \bar{g}(x) = \overline{g^{**}}(x) = g^{**}(\det \nabla u_\infty(x))$ a.e. *in* Ω;
2. $\operatorname{supp} \mu_x \subset \{M \in M_+^{3\times 3} | \alpha(x) \le \det M \le \beta(x)\}$ *where* $[\alpha(x), \beta(x)]$ *is the maximal closed interval containing* $\det \nabla u_\infty(x)$ *on which* g^{**} *is affine;*
3. $W(M) = g(\det M) = g^{**}(\det M)$ a.e. *in* $\operatorname{supp} \mu_x$.

Proof. The argument is essentially the same as in CHIPOT & KINDERLEHRER [2], where we must use Proposition 2.9 (1).

Finally, we give the proof of Theorem 3.1.

Proof of Theorem 3.1. By Proposition 3.6 (3), if $M \in \operatorname{supp} \mu_x$ then

$$0 = W(M) - g^{**}(\det M) = \min\{W(.) - g^{**}(\det .)\}$$

and so

$$0 = \frac{\partial W}{\partial M}(M) - (g^{**})'(\det M)\,\mathrm{adj}\ M.$$

Also, by Proposition 3.6 (2)

$$(g^{**})'(\det M) = (g^{**})'(\det \nabla u_\infty(x))$$

and we deduce that

$$S(M) = (g^{**})'(\det \nabla u_\infty(x))\,\mathrm{adj}\ M, \tag{3.5}$$

hence, by Proposition 2.9 (2),

$$\begin{aligned} \overline{S}(x) &= \int_{M^{3\times 3}} (g^{**})'(\det \nabla u_\infty(x))\,\mathrm{adj}\ M d\mu_x(L) \\ &= (g^{**})'(\det \nabla u_\infty(x))\,\mathrm{adj}\ L_\infty(x). \end{aligned} \tag{3.6}$$

Finally, by (3.5) and for almost all $x \in \Omega$

$$\begin{aligned} \overline{T}(x) &= \int_{M^{3\times 3}} T(M) d\mu_x(L) = \int_{M^{3\times 3}} \frac{1}{\det M} S(M) M^T d\mu_x(M) \\ &= (g^{**})'(\det \nabla u_\infty(x)) \int_{M^{3\times 3}} \frac{1}{\det M}\,\mathrm{adj}\ M M^T d\mu_x(M) \\ &= (g^{**})'(\det \nabla u_\infty(x))\mathbf{1} \end{aligned}$$

and by (3.6)

$$\begin{aligned} \frac{1}{\det L_\infty(x)} \overline{S}(x) L_\infty^T &= \frac{1}{\det L_\infty(x)} (g^{**})'(\det \nabla u_\infty(x))\,\mathrm{adj}\ L_\infty(x) L_\infty^T \\ &= (g^{**})'(\det \nabla u_\infty(x))\mathbf{1}. \end{aligned}$$

4. A variational problem involving bulk and interfacial energies. As shown in Theorem 3.1, including neutrally related states as admissible changes seems to render the material weak as it cannot sustain nonzero average shear stresses. So we contemplate problems where the slip, represented by v, is penalized. Now, minimizing $E(.,.)$ corresponds, formally, to a variational problem involving variational of the domain. Indeed, if v was an invertible mapping then

$$I(\Sigma) = \int_{v(\Omega)} W(\nabla\omega(y))dy$$

where

$$\omega(y) := u \circ v^{-1}(y).$$

If the solid attained equilibrium at an elastic state then $v(\Omega)$ would be a translation of Ω, i.e. $v(x) = x+$ Const. In this section, we will consider an energy functional were we penalize the boundary $v(\Omega)$ in order to obtain some information regarding the preferred shape for $v(\Omega)$.

Suppose that $u_0 \in C^1(\overline{\Omega}; \mathbb{R}^3)$ is one-to-one in Ω and $\det \nabla u_0(x) > 0$ in Ω. Suppose further that

$$\int_\Omega g^{**}(\det \nabla u_0(x))dx \leq \int_\Omega g^{**}(\det \nabla u(x))dx \tag{4.1}$$

for all $u \in u_0 + W^{1,\infty}(\Omega; \mathbb{R}^3)$. Note that, by Jensen's inequality and using the fact that $F \to \det F$ is a null lagrangian, (4.1) is trivially satisfied if $\det \nabla u_0 =$ constant. Consider the energy functional

$$\tilde{I}(\Sigma) = \tilde{E}(u,v) := \int_\Omega W(\nabla u(x)\{\nabla v(x)\}^{-1})dx + \int_{\partial^* v(\Omega)} \Gamma(n(x))dH_2(x)$$

where $(u,v) \in \mathcal{A}^*(u_0) := \{(u,v) \in W^{1,\infty}(\Omega, \mathbb{R}^3) | \det \nabla u > 0$ a.e. in $\Omega, u = u_0$ on $\partial\Omega$, $\det \nabla v = 1$ a.e. in Ω, meas $v(\Omega) =$ meas $\Omega\}$, $\partial^* v(\Omega)$ denotes the reduced boundary of $v(\Omega)$, $n(x)$ is the outer unit normal to the boundary of $v(\Omega)$, Γ is a Lipschitz (anisotropic) surface energy density and H_2 is the two-dimensional Hausdorff measure.

Remark 4.1. $\tilde{I}(\Sigma)$ does not depend on the representation (u,v) of Σ. Indeed, it can be shown that

$$(\tilde{u}, \tilde{v}) \text{ is another representation of } \Sigma$$

if and only if

$$v(x) = \tilde{v}(f(x)) + \text{ Cont. a.e. in } \Omega.$$

where

$$f(x) = x \text{ on } \partial\Omega \text{ and } \det \nabla f(x) = 1 \text{ a.e. in } \Omega.$$

Hence,

$$v(\Omega) \text{ is a translation of } \tilde{v}(\Omega)$$

and so

$$\widetilde{E}(u,v) = \widetilde{E}(\tilde{u},\tilde{v}).$$

Also, by the definition of $\mathcal{A}^*(u_0)$, v is subject to the constraints $\det \nabla v = 1$ a.e. in Ω and $\text{meas}\, v(\Omega) = \text{meas}\,(\Omega)$. These conditions imply that v is invertible a.e. in Ω and $v^{-1} \in W^{1,\infty}(\Omega;\mathbf{R}^3)$ (see BALL [1]).

It is well known (see HERRING [14], FONSECA [10], FONSECA & MÜLLER [11], TAYLOR [18], [19], [20], WULFF [21]) that the shape that minimizes the surface energy

$$\int_{\partial^* A} \Gamma(n)dH_2$$

among all sets A of finite perimeter with constant volume, meas A = meas Ω, is the *Wulff set*

$$C := \{x \in \mathbf{R}^3 | x.n \le \gamma(n) \text{ for all unit vector } n\}^4.$$

Here, we assume that $(H)\Omega$ is a bounded, open, star-shaped domain[4] with respect to $x_0 \in$ int Ω, $B(x_0,\varepsilon) \subset \Omega$ for some $\varepsilon > 0$, and there exists a finite partition of Ω, $\Omega = \Omega_1 \cup \Omega_2 \cup \cdots \cup \Omega_p$, such that Ω_i is a cone with vertex at x_0, $\partial(\Omega_i) \cap \partial\Omega \in C^1$, $B(x_0,\varepsilon) \cap \Omega_i$ is convex for all $i = 1,\ldots,p$. Also, there exist $\alpha > 0$ such that $n(x).x \ge \alpha$ for all $x \in \partial\Omega$, where $n(x)$ denotes the outward unit normal to $\partial\Omega$ at x.

THEOREM 4.2. *Let $u_0 \in C^1(\overline{\Omega};\mathbf{R}^3)$, be such that $\det \nabla u_0(x) > 0$ in Ω and (4.1) holds. If C and Ω satisfy the condition (H)[6] then*

$$\inf\{\widetilde{E}(u,v) | (u,v) \in \mathcal{A}^*(u_0)\} = \int_\Omega g^{**}(\det \nabla u_0(x))dx + \int_{\partial^* C} \Gamma(n(x))dH_2(x).$$

The proof of Theorem 4.2 relies on the following two results

PROPOSITION 4.3. *If $y_0 \in C^1(\overline{\Omega};\mathbf{R}^3)$ is such that $\det \nabla y_0(x) > 0$ in Ω and if $y \in C^1(\overline{\Omega};\mathbf{R}^3)$ verifies $y = y_0$ on $\partial\Omega$ then*

$$\inf\left\{\int_\Omega W(\nabla u(x))dx | u \in y_0 + W^{1,\infty}(\Omega;\mathbf{R}^3)\right\} \le \int_\Omega g^{**}(\det \nabla y(x))dx.$$

Proof. It follows immediately from the proof of Theorem 4.1 in CHIPOT & KINDERLEHRER [3].

---

[4] Here, and for simplicity, we suppose that meas Ω = meas C.

[4] We say that Ω is *star-shaped* with respect to an interior point $x_0 \in \Omega$ if any ray with origin at x_0 has a unique common point with the boundary $\partial\Omega$.

[6] In general, the Wulff set C for solid crystals is a polyhedron.

THEOREM 4.4. *If Ω and Ω' satisfy (H) and if* $\text{meas}(\Omega) = \text{meas}(\Omega')$ *then there exists $v \in W^{1,\infty}(\Omega, \Omega')$ such that $v(\Omega) = \Omega'$,* $\det \nabla v = 1$ a.e. *in Ω, v is invertible and $v \in C^1(\overline{U}_i)$ for some finite partition of Ω into strongly Lipschitz domains $\{U_i\}$.*

The proof of Theorem 4.4 can be found in FONSECA & PARRY [12].

Proof of Theorem 4.2. By (4.1) and Proposition 3.3 we have

$$
\begin{aligned}
\widetilde{E}(u,v) := & \int_{\Omega} W(\nabla u(x)\{\nabla v(x)\}^{-1})dx + \int_{\partial^* v(\Omega)} \Gamma(n(x))dH_2(x) \\
\geq & \int_{\Omega} g^{**}(\det \nabla u_0(x))dx + \int_{\partial^* C} \Gamma(n(x))dH_2(x).
\end{aligned}
\tag{4.2}
$$

On the other hand, using Theorem 4.4 let $v \in W^{1,\infty}(\Omega, C)$ be such that $v(\Omega) = C$, $\det \nabla v = 1$ a.e. in Ω, v is invertible and $v \in C^1(\overline{\Omega}_i)$ for some finite partition of Ω into strongly Lipschitz domains $\{\Omega_i\}_{i=1,\dots p}$. Clearly, $u_0 \circ v^{-1} \in C^1(\overline{v(\Omega_i)})$ and so, by Proposition 4.3 and changing variables

$$
\begin{aligned}
\inf & \left\{ \int_{v(\Omega_i)} W(\nabla w(z))dz | w \in W^{1,\infty}(v(\Omega_i)), \det \nabla w > 0 \text{ a.e. }, w = u_0 \circ v^{-1} \text{ on } \partial v(\Omega_i) \right\} \\
& \leq \int_{v(\Omega_i)} g^{**}(\det \nabla (u_0 \circ v^{-1})(z))dz \\
& = \int_{\Omega_i} g^{**}(\det \nabla u_0(x))dx.
\end{aligned}
$$

Fix $\varepsilon > 0$ and let $w_i \in W^{1,\infty}(v(\Omega_i))$, be such that $\det \nabla w > 0$ a.e., $w = u_0 \circ v^{-1}$ on $\partial v(\Omega_i)$ and

$$
\int_{v(\Omega_i)} W(\nabla w_i(z))dz \leq \int_{\Omega_i} g^{**}(\det \nabla u_0(x))dx + \varepsilon.
\tag{4.3}
$$

Let $w(z) = \sum_{i=1}^p \chi_{v(\Omega_i)}(z) w_i(z)$ if $z \in C$, where χ_A denotes the characteristic function of the set A. Clearly $w \in W^{1,\infty}(C)$, $\det \nabla w > 0$ a.e. in Ω, $w = u_0 \circ v^{-1}$ on ∂C and by (4.3)

$$
\begin{aligned}
\int_C W(\nabla w(z))dz &= \sum_{i=1}^p \int_{v(\Omega_i)} W(\nabla w_i(z))dz \\
&\leq \sum_{i=1}^p \int_{\Omega_i} g^{**}(\det \nabla u_0(x))dx + p\varepsilon \\
&= \int_{\Omega} g^{**}(\det \nabla u_0(x))dx + p\varepsilon.
\end{aligned}
$$

Thus, letting $\varepsilon \to 0^+$ we deduce that
(4.4)

$$\inf\left\{\int_{v(\Omega)} W(\nabla w(z))dz | w \in W^{1,\infty}(v(\Omega)), \det \nabla w > 0 \text{ a.e. and } w = u_0 \circ v^{-1} \text{ on } \partial v(\Omega)\right\}$$
$$\leq \int_\Omega g^{**}(\det \nabla u_0(x))dx$$

and, by (4.2) and changing variables we have with $u = w \circ v$

$$\begin{aligned}\int_{v(\Omega)} W(\nabla w(z))dz &= \int_\Omega W(\nabla w(v(x)))dx \\ &= \int_\Omega W(\nabla u(x)\{\nabla v(x)\}^{-1})dx \\ &\geq \int_\Omega g^{**}(\det \nabla u_0(x))dx\end{aligned}$$

which, together with (4.4) yields

$$\inf\left\{\int_{v(\Omega)} W(\nabla w(z))dz | w \in W^{1,\infty}(v(\Omega)), \det \nabla w > 0 \text{ a.e. and } w = u_0 \circ v^{-1} \text{ on } \partial v(\Omega)\right\}$$
$$= \int_\Omega g^{**}(\det \nabla u_0(x))dx.$$

Let $w_k \in W^{1,\infty}(v(\Omega))$ be such that $\det \nabla w_k > 0$ a.e., $w_k = u_0 \circ v^{-1}$ on $\partial v(\Omega)$ and

$$\int_{v(\Omega)} W(\nabla w_k(z))dx \to \int_\Omega g^{**}(\det \nabla u_0(x))dx$$

and set $u_k := w_k \circ v$. We conclude that

$$\begin{aligned}\widetilde{E}(u_k, v) :&= \int_\Omega W(\nabla u_k(x)\{\nabla v(x)\}^{-1})dx + \int_{\partial^* v(\Omega)} \Gamma(n(x))dH_2(x) \\ &= \int_{v(\Omega)} W(\nabla w_k(z))dz + \int_{\partial^* C} \Gamma(n(x))dH_2(x) \\ &\to \int_\Omega g^{**}(\det \nabla u_0(x))dx + \int_{\partial^* C} \Gamma(n(x))dH_2(x)\end{aligned}$$

and by (4.2) the result is proven.

REFERENCES

[1] BALL, J.M., *Global invertibility of Sobolev functions and the interpenetration of matter*, Proc. Roy. Soc. Edinburgh 88A (1981), pp. 315–328.

[2] CHIPOT, M. AND D. KINDERLEHRER, *Equilibrium configurations of crystals*, Arch. Rat. Mech. Anal. 103 (1988), pp. 237–277.

[3] DAVINI, C., *A proposal for a continuum theory of defective crystals*, Arch. Rat. Mech. Anal. 96 (1986), pp. 295–317.

[4] DAVINI, C. AND G. PARRY, *On defect-preserving deformations in crystals*, Int. J. of Plasticity 5 (1989), pp. 337–369.

[5] DAVINI, C. AND G. PARRY, *A complete list of invariants for defective crystals*, to appear.

[6] ERICKSEN, J.L., *Loading devices and stability of equilibrium*, in Nonlinear Elasticity, Academic Press, N.Y. (1973), pp. 161–173.

[7] ERICKSEN, J.L., *Twinning of crystals I*, in Metastability and Incompletely Posed Problems, IMA Vol. in Appl. Math 3 (1987), pp. 77–96.

[8] FONSECA, I., *Variational methods for elastic crystals*, Arch. Rat. Mech. Anal. 97 (1987), pp. 189–220.

[9] FONSECA, I., *The lower quasiconvex envelope of the stored energy function for an elastic crystal*, J. Math. Pures et Appl. 67 (1988), pp. 175–195.

[10] FONSECA, I., *The Wulff Theorem revisited*, Proc. R. Soc. Lond. A 432 (1991), pp. 125–145.

[11] FONSECA, I. AND S. MÜLLER, *An uniqueness proof for the Wulff Problem*, Proc. Roy. Soc. Edinburgh 119 A (1991), pp. 125–136.

[12] FONSECA, I. AND G. PARRY, *Equilibrium configurations of defective crystals*, to appear.

[13] FONSECA, I. AND G. PARRY, *On a class of invariant integrals*, to appear.

[14] HERRING, C., *Some theorems on the free energies of crystal surfaces*, Phys. Rev. 82 (1951), pp. 87–93.

[15] KINDERLEHRER, D., *Twinning of crystals II*, in Metastability and Incompletely Posed Problems, IMA Vol. in Appl. Math 3 (1987), pp. 185–211.

[16] TARTAR, L., *Compensated compactness and applications to partial differential equations*, Nonlinear analysis and mechanics: Heriot-Watt Symposium, vol. IV, Res. Notes in Math. 39, Pitman, 1979, pp. 136–212.

[17] TAYLOR, G.I., *The mechanism of plastic deformation of crystals, I,II*, Proc. Roy. Soc. A 145 (1934), pp. 362–387, 388–404.

[18] TAYLOR, J., *Existence and structure of solutions to a class of nonelliptic variational problems*, Symposia Mathematica 14 (1974), pp. 499–508.

[19] TAYLOR, J., *Unique structure of solutions to a class of nonelliptic variational problems*, Proc. Symp. Pure Math., A.M.S., 27 (1975), pp. 419–427.

[20] TAYLOR, J., *Crystalline variational problems*, Bull. Amer. Math. Soc. 84 (1978), pp. 568–588.

[21] WULFF, G., *Zur Frage der Geschwindigkeit des Wachstums und der Auflösung der Kristallflächen*, Zeitschrift für Kristallographie 34 (1901), pp. 449–530.

[22] YOUNG, L.C., *Generalized surfaces in the calculus of variations I,II*, Ann. of Math. 43 (1942), pp. 84–103, 530–544.

THE CONCEPT OF A LAMINATE

PABLO PEDREGAL†

Abstract. In connection with phase transitions problems in crystals, the term laminate has been successfully used to describe the behavior of such crystals under some special external loads. We will attempt to give this term a precise mathematical significance and explore its relationship with rank-one convexity, quasiconvexity and Young measures. Roughly speaking, laminates are to rank-one convexity what Young measures are to quasiconvexity.

Key words. (H_n) conditions, laminate, rank-one convexity.

AMS(MOS) subject classifications. 73C50, 73C60, 73S99

1. Introduction. Many people is now familiar with the type of situation that may occur in the study of certain variational principles in elasticity which lack weak lower semicontinuity, often as a consequence of some specific material symmetry. In these circumstances a minimizing sequence may oscillate wildly between two or more homogeneous states. To describe heuristically such phenomenon, the term "laminate" has been used, which by itself provides an intuitive idea of the kind of behavior we might expect. We also talk about "microstructure" referring to the fine inner structure that such materials exhibit in an attempt to achieve the minimum energy available to it. Thus, laminate is a descriptive term for microstructure. What we seek here is to provide a mathematical framework to deal with microstructure in all its complexity as it arises in the analytical description of phase transitions in crystalline solids. The literature on this topic is abundant ([1-4], [13-20], [23-25], [31], [32]) and we will not insist on this. We simply outline the most basic example of a laminate as our starting point and build up on this.

The idea comes from the concept of (H_n) condition, originally introduced by B. Dacorogna in 1985 ([11]), and thereafter we complete the set of such probability measures by taking weak * limits of such objects. The resulting family of probability measures is the family of laminates and the underlying structure of these measures is revealed with the graphical interpretation we give of the (H_n) conditions in terms of tree-structures which prove to be quite convenient for our purposes. One important feature is that laminates can be realized as actual deformation gradients or limit states of such deformations. In fact, laminates is a very special class of gradient Young measures, and all of the explicit examples we are aware of at the moment are laminates by construction or there is no known way to decide what they are. This difficulty is closely related to the problem of rank-one convexity versus quasiconvexity as we will point out later. Some of the nice facts one can obtain with laminates are due to their inner structure, something that gradients, at first sight, lack. It would be quite remarkable that gradients shared this same structure as a consequence of the equivalence of those two notions of convexity.

†Departamento de Matemática Aplicada, Universidad Complutense, 28040 Madrid, Spain.

The motivation to go through this enterprise comes from several sources. In the first place, a rather striking example whose genesis goes back to the early 80's when J. Ball, R. Kohn and L. Tartar were discussing about separately convex functions. It was later rediscovered by V. Nesi and G. Milton as a non-trivial example of an infinite-levelled microstructure in which only a finite number of matrices participate. We will precisely explain what we mean by such a microstructure or laminate. Secondly, it turns out that this general concept of laminate as will be presented here can be characterized by Jensen's inequality for rank-one convex functions. The importance of this fact becomes apparent after the analogous characterization for Young measures by quasiconvex functions proved by D. Kinderlehrer and P. Pedregal [26], so that laminates are to rank-one convexity what Young measures are to quasiconvexity. And therefore the distinction between Young measures and laminates is as subtle as the distinction between rank-one convexity and quasiconvexity. How important this approach is regarding this classical and fundamental problem is not clear yet. But the more we learn about laminates the closer we are to discover similarities or differences between these two classes of objects. And finally, looking at the two-well problem in two dimensions, D. Kinderlehrer and P. Pedregal ([27]) found continuously distributed microstructures. These require an infinite process in order to be described and can be very well understood in the context we are about to introduce. Physically, these highly complex microstructure have not been observed. Nor have they been obtained numerically. From either point of view it is hard to get "close" to infinite fineness, for obvious reasons, so that an infinite number of matrices might show up in the microstructure. No more than four matrices participate in the microstructures observed experimentally or numerically, though it has been suggested that some computations by C. Collins and M. Luskin ([7-9]) seem to tend to a microstructure supported in seven different matrices.

We end with some remarks on the three-matrices case. This has been studied by R. Kohn ([29], [30]) and N. Firoozye ([21], [22]) using some algebraic ideas which were completely successful for the two-matrices case (see also Pipkin [35]), but so far not so good when we add another matrix. We concentrate here in determining whether there can be a non-trivial laminate supported in three different matrices, not pairwise rank-one related.

This paper is essentially expository in nature for the proofs of the different results are rather technical if not hard and we omit them. For more precise information we refer to [33] and [34].

I wish to take this opportunity to thank the constant support of the research group on *Transitions and Defects in Ordered Materials* and the IMA. They both made possible my assistance and my participation in the workshop. I am most grateful to them all.

Let us review very briefly how these special minimizing sequences come up in elasticity. We follow here Chipot and Kinderlehrer [6]. Assume we have a crystal with stored energy function W such that

$$W \geq 0, \quad W(\mathbf{1}) = 0.$$

The hypothesis of frame indifference and internal symmetry of the crystal translate into the invariance

$$W(F) = W(RFH)$$

for any $R \in SO(3)$ and $H \in \mathbf{H}$, an appropriate subgroup of matrices related to the symmetry properties of the crystal.

Let us further assume that $H = \mathbf{1} + a \otimes n \in \mathbf{H}$, where $a \otimes n$ is the matrix $(a \otimes n)_{ij} = a_i n_j$, a, $n \in \mathbf{R}^3$, $F = \mathbf{1} + \theta a \otimes n \notin \mathbf{H}$, $\theta \in (0,1)$ and look at the variational principle

$$m = \inf_{\mathcal{A}} \int_\Omega W(\nabla u)\, dx$$

where Ω is a bounded, regular domain with measure 1 and

$$\mathcal{A} = \{ v \in H^{1,\infty}(\Omega) : v(x) = F \cdot x, \quad x \in \partial\Omega \}$$

for $F = \mathbf{1} + \theta a \otimes n = (1-\theta)\mathbf{1} + \theta H$, $\theta \in (0,1)$. Then we claim that $m = 0$ even though $W(F) > 0$.

To show this, set $y(x) = Fx$ and let χ be the characteristic function of $(0, \theta)$ in $(0,1)$ extended periodically to all of $\mathbf{R}$. Write then

$$f^k(x) = \chi(k n \cdot x), \qquad k = 1, 2, 3, \ldots$$

so that it is well known that $f^k \overset{*}{\rightharpoonup} \theta$ in $L^\infty(\Omega)$ for any bounded $\Omega \subset \mathbf{R}^3$. Let

$$u(\xi) = \xi + \int_0^{\xi \cdot n} \chi(t)\, dt \cdot a$$

and

$$u^k(x) = k^{-1} u(kx),$$

so that

$$\nabla u^k = (1 - f^k)\mathbf{1} + f^k H.$$

Then $u^k \overset{*}{\rightharpoonup} y$ in $H^{1,\infty}(\Omega)$ which in particular means $u^k \to y$ uniformly in Ω. Using this fact one can arrange the boundary values for u^k. Take η^k, a cut-off function with the properties

i) $\eta^k = 0$ on $\partial\Omega$,

ii) $\eta^k = 1$ in $\Omega_k = \{x \in \Omega : \operatorname{dist}(x, \partial\Omega) \geq 1/k\}$,

iii) $|\nabla u^k| \leq Ck$.

Then

$$v^k = \eta^{j_k} u^k + (1 - \eta^{j_k}) y$$

is admissible, $\|\nabla v^k\|_{L^\infty(\Omega)} \leq M < \infty$ for suitable $\{j_k\}$ and $v^k = u^k$ in Ω_k. We also have $v^k \overset{*}{\rightharpoonup} y$ in $H^{1,\infty}(\Omega)$. But one checks very easily

$$\int_\Omega W(\nabla v^k)\, dx \to 0$$

even in the case

$$\int_\Omega W(\nabla y)\,dx > 0.$$

So the energy functional is not weak lower semicontinuous. But what we want to emphasize in this construction is the special structure of ∇u^k which takes the values $\mathbf{1}$ and H in alternate layers of relative width $1-\theta$ and θ. One final observation is that the Young measure associated with ∇u^k or ∇v^k is $\nu = (1-\theta)\delta_{\mathbf{1}} + \theta\delta_H$ as an analogous computation reveals:

$$\lim_{k\to\infty} \int_E \varphi(\nabla u^k)\,dx = \int_E \int_{\mathbf{M}} \varphi(A)\,d\nu(A)\,dx,$$

for any continuous φ and any measurable E.

Let us go for a moment one step further. We now have three matrices $\mathbf{1}$, F^+ and F^- such that

$$\begin{aligned} \mathbf{1} - F &= \mathbf{1} - \big((1-\lambda)F^+ + \lambda F^-\big) = a \otimes n, \\ F^+ - F^- &= b \otimes m. \end{aligned}$$

We claim, then, that we can find a sequence of gradients, ∇u^k, with boundary values $[(1-\theta)\mathbf{1} + \theta F]\,x$, $x \in \partial\Omega$ and such that for any continuous $\varphi : \mathbf{M} \to \mathbf{R}$,

$$\int_E \varphi(\nabla u^k)\,dx \to \int_E \int_{\mathbf{M}} \varphi(A)\,d\nu(A)\,dx$$

for any $E \subset \Omega$ and this time $\nu = (1-\theta)\delta_{\mathbf{1}} + \theta(1-\lambda)\delta_{F^+} + \theta\lambda\delta_{F^-}$. It is not difficult to infer that, roughly speaking, one would get ∇u^k putting layers within layers, that is to say, in two steps like the one we did. And this can be done because of the kinematically compatibility conditions above (see Figure 1).

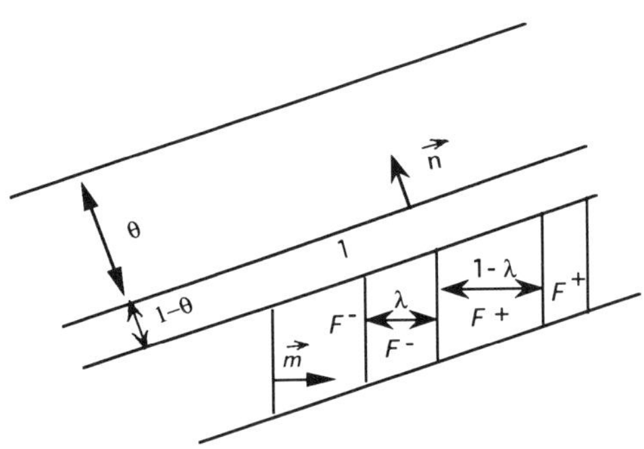

Figure 1

2. (H_n) conditions and laminates. It is not surprising how one can generalize the above situation when a finite number of matrices is involved. All we need to carry out that construction is the compatibility conditions expressed in those

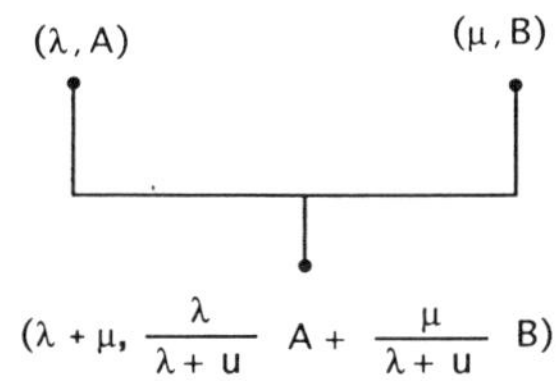

Figure 2

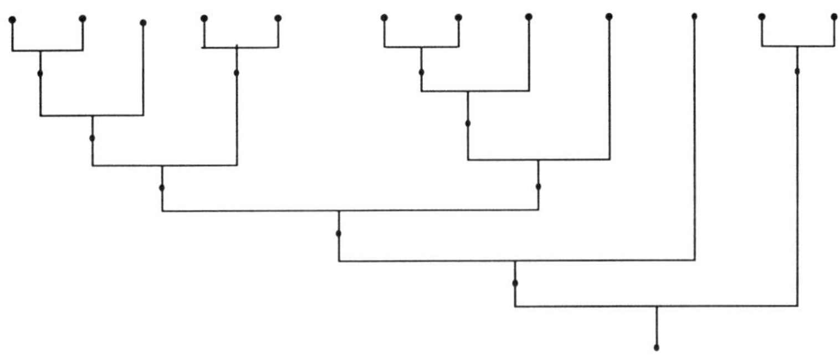

Figure 3

rank-one connections. And here we borrow the concept of (H_n) condition from Dacorogna [11] or [12], for this is exactly what we need.

DEFINITION. *The set of pairs* $\{(\lambda_i,\, A_i)\}_{1\le i\le n}$, $\sum \lambda_i = 1$, $\lambda_i > 0$ *and* A_i *are matrices, verifies the* (H_n) *condition if:*

i) For $n = 2$: $\operatorname{rank}(A_1 - A_2) \le 1$.

ii) For $n > 2$, *and possibly after a permutation of indices,*

$$\operatorname{rank}(A_1 - A_2) \le 1$$

and if we set

$$\mu_1 = \lambda_1 + \lambda_2,$$
$$B_1 = \frac{\lambda_1}{\mu_1} A_1 + \frac{\lambda_2}{\mu_2} A_2,$$
$$\mu_i = \lambda_{i+1}, \quad B_i = A_{i+1}, \quad 2 \le i \le n-1,$$

the set of pairs $\{(\mu_i,\, B_i)\}_{1\le i\le n-1}$ *satisfies* (H_{n-1}).

Instead of discussing this definition analytically, it is much better to give a graphical explanation that is very convenient and clear. Let us agree to represent the pairs $(\lambda_i,\, A_i)$ by black dots and the transformation in ii) of the definition as shown in Figure 2. Then any (H_n) condition can be represented in a tree-like fashion like the one in Figure 4.

One important observation here is that different dots do not necessarily correspond to different pairs. So we see how the compatibility conditions are adjusted so that at the end we can reach the underlying deformation $F = \sum \lambda_i A_i$. And these compatibility conditions or rank-one connections allow us to actually construct the corresponding sequence of gradients with the right properties.

Another observation is that given the initial set of pairs $\{(\lambda_i, A_i)\}_{1 \le i \le n}$ the tree diagram might not be unique.

We are now ready for the following definition.

DEFINITION. *Given a compact set of matrices $\mathcal{K}$, and $F \in co(\mathcal{K})$, we call a laminate supported in $\mathcal{K}$ with underlying deformation F, a probability measure ν in $\mathcal{K}$ with barycenter F and such that there exists a sequence of sets of pairs $\{(\lambda_i^k, A_i^k)\}_{1 \le i \le n_k}$ verifying the (H_{n_k}) condition and*

i) for every continuous function $\varphi : \mathbf{M} \to \mathbf{R}$

$$< \nu, \varphi >= \lim_{k \to \infty} \sum_i \lambda_i^k \varphi \left(A_i^k\right);$$

ii) for every k,

$$F = \sum_i \lambda_i^k A_i^k.$$

It is not difficult to deduce from i) above:

iii) $|A_i^k| \le M$ uniformly in i, k;

iv) $\forall U$, open neighborhood of $\mathcal{K}$,

$$\lim_{k \to \infty} \sum_{A_i^k \notin U} \lambda_i^k = 0.$$

So therefore a laminate is a weak * limit in $\mathcal{M}(\mathcal{K})$ (the set of measures in $\mathcal{K}$) of probability measures associated to (H_n) conditions. And thus it should be clear (keeping in mind the uniform bound iii))that any laminate is a gradient Young measure, that is to say, we can find a sequence of deformations $\{u^k\}$ with the following requirements:

1) $u^k(x) = y(x)$, $x \in \partial\Omega$, where $y(x) = Fx$ and $F = \int_{\mathcal{K}} A \, d\nu(A)$;

2) $u^k \stackrel{*}{\rightharpoonup} y$ in $H^{1,\infty}(\Omega)$;

3)
$$\lim_{k \to \infty} \int_E \varphi\left(\nabla u^k\right) dx = \int_E \int_{\mathcal{K}} \varphi(A) \, d\nu(A) \, dx = |E| \int_{\mathcal{K}} \varphi(A) \, d\nu(A),$$
for any measurable E and continuous φ.

3. Examples. Let us consider some examples. The simplest one is $\mathcal{K} = \{A, B\}$. In this case we know all the laminates: if rank$(A - B) > 1$ then the only laminates are δ_A and δ_B, but if rank$(A - B) = 1$ then any convex combination

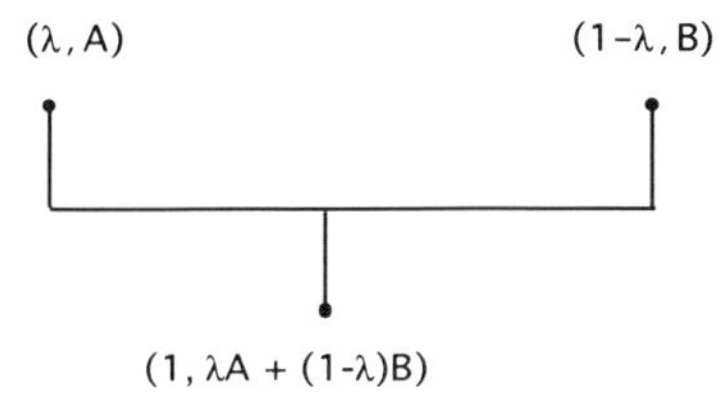

Figure 4

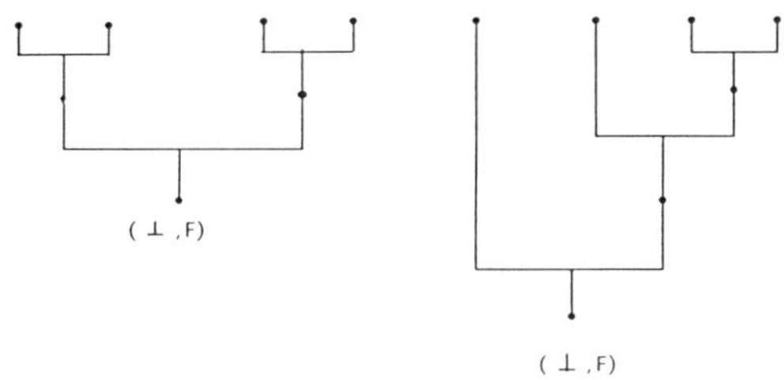

Figure 5

of those two delta measures is a laminate. This last remark is clear, for then the tree-like structure is simply

But the first statement is not so straightforward. There might be some special way of arranging the (H_{n_k}) conditions so that we would get that $\nu = \lambda\delta_A + (1-\lambda)\delta_B$ is a laminate. That this is not the case is a consequence of using the weak * continuity of the 2×2-minors; that is, if M represents such a minor, then we should have

$$\int_K M(P)\, d\nu(P) = M\left(\int_K P\, d\nu(P)\right)$$

or

$$\lambda M(A) + (1-\lambda)M(B) = M\left(\lambda A + (1-\lambda)B\right),$$

and this forces $\operatorname{rank}(A-B) \leq 1$.

We will discuss the case of three matrices at the end, although it can be guessed at this point the easy examples when there are rank-one connections joining the matrices. With four matrices, we want to look first at the straightforward examples. Figure 5 shows them in terms of diagrams and Figure 6, geometrically (lines represent rank-one directions).

But there is also a different situation that it cannot be achieved in a finite number of steps. We are referring to the clever example mentioned in the introduction. Geometrically, it is the one shown in Figure 7. For the F in this picture one would have the sequence of (H_n) conditions in Figure 8. So it is clear that F is supported by a laminate.

This particular configuration is the one Ball, Kohn and Tartar were reflecting upon from the point of view of separately convex functions of two variables. Specifically, they realized that if φ is separately convex then we can say something about $\varphi(F)$ in terms of $\varphi(A_i)$. In the language of laminates this is nothing other than

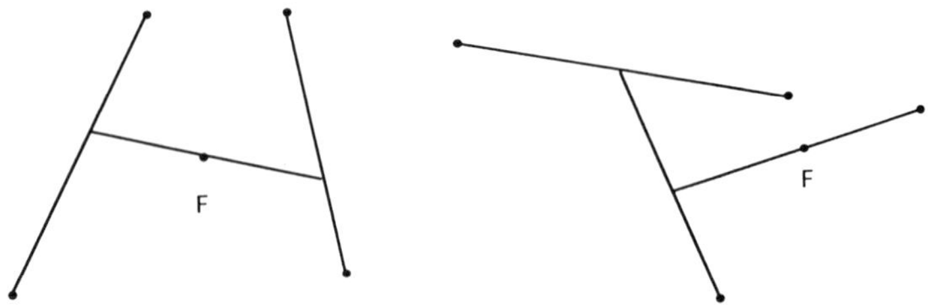

Figure 6

Jensen's inequality

$$\varphi\left(\int_{\mathcal{K}} A\, d\nu\right) = \varphi(F) \leq \int_{\mathcal{K}} \varphi(A)\, d\nu$$

for $\mathcal{K} = \{A_1, A_2, A_3, A_4\}$ and ν the laminate just described. We will come back to this type of questions in the next section. And once this example is understood, other similar configurations can be figured out with four matrices spanning a three-dimensional linear manifold or even examples with more than four matrices. In this sense, we will show in the final section that this sort of example requires at least four matrices and there is nothing of the type with just three matrices.

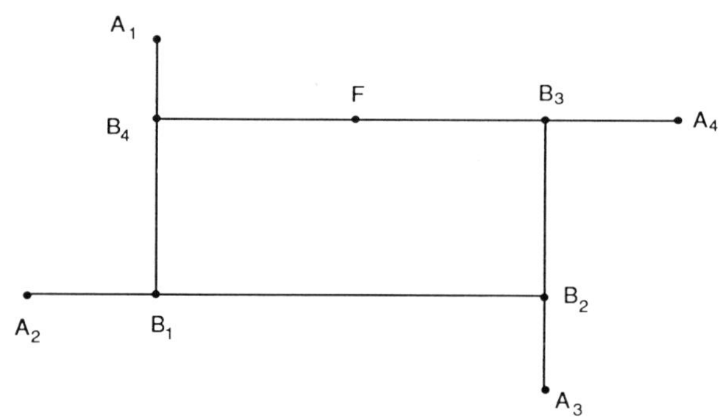

Figure 7

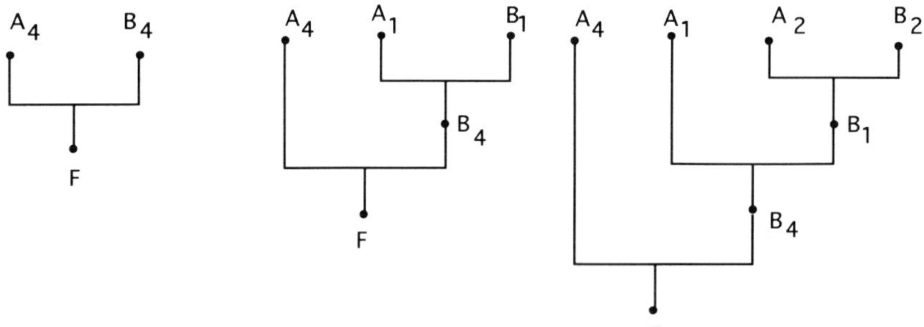

Figure 8

For applications to more physically real problems, the compact set of matrices $\mathcal{K}$ is the zero set of the stored energy function or set of potential wells. This set $\mathcal{K}$ should incorporate the invariance properties relative to material symmetry. Then it is elementary to check that a laminate minimizer (a laminate associated to a minimizing sequence) with zero energy is a laminate whose support lies in $\mathcal{K}$, and so the admissible microstructures are exactly the set of laminates with support in $\mathcal{K}$. In general, finding all these microstructures is a pretty hard problem. Special cases can be found through the references, fruit of many people's work. Of particular interest, D. Kinderlehrer and P. Pedregal ([26-28]) have studied general properties of sets of microstructures and, as announced, even found continuously distributed laminates when $\mathcal{K}$ is the set of two potential wells (see also [5], [10]).

4. Characterization by rank-one convex functions. There is no doubt that rank-one convexity might have a close connection with laminates, since a crucial ingredient of a laminate is the rank-one compatibility relations. Suppose that we have a rank-one convex function φ which is continuous, and a family of pairs $\{(\lambda_i, A_i)\}_{1\leq i\leq n}$ satisfying the (H_n) condition. Then, it is pretty clear because of the recursive way (H_n) conditions are defined that

$$\varphi\left(\sum_i \lambda_i A_i\right) = \varphi(F) \leq \sum_i \lambda_i \varphi(A_i),$$

which is nothing other that Jensen's inequality for the laminate $\sum \lambda_i \delta_{A_i}$ as we already noticed. If now ν is a laminate according to our general definition, and because of the continuity of φ, then

$$\varphi\left(\int_{\mathcal{K}} A\, d\nu\right) \leq \int_{\mathcal{K}} \varphi(A)\, d\nu$$

which again is Jensen's inequality. So this type of inequality is a necessary condition for any rank-one convex, continuous function and any laminate. What is rather interesting is that this is exactly a characterization of laminates according to our definition. More precisely,

THEOREM. *A probability measure ν, supported in $\mathcal{K}$ with barycenter*

$$F = \int_{\mathcal{K}} A\, d\nu(A) \in co(\mathcal{K})$$

is a laminate if and only if for every rank-one convex, continuous function φ, Jensen's inequality holds:

$$\varphi(F) \leq \int_{\mathcal{K}} \varphi(A)\, d\nu(A).$$

Comparison with the corresponding theorem in [26] reveals what we already pointed out in the introduction: this theorem remains true word by word changing laminate by Young measure and rank-one convex by quasiconvex. Thus, if we could find a Young measure ν with compact support which is not a laminate then, by these characterizations, there should be at least one rank-one convex, continuous function φ_0 for which

$$\varphi_0(F) > \int_{\mathcal{K}} \varphi_0(A)\, d\nu(A).$$

But since ν is a Young measure, φ_0 can not be rank-one quasiconvex. And conversely, if laminates and Young measures are the same thing (regardless of the support $\mathcal{K}$) then rank-one convexity and quasiconvexity are equivalent concepts. This is not so clear from these Jensen's inequalities. But for any $H^{1,\infty}(\Omega)$-function, u with affine boundary value F, we can talk about the average Young measure $\nu = \overline{\delta_{\nabla u}}$ (as defined in [26]) obtained by homogenization, which is itself an homogeneous Young measure, and therefore a laminate if we assume they are the same. This particular laminate has compact support and underlying deformation

$$\int_{\mathbf{M}} A\, d\nu(A) = \int_{\Omega} \nabla u\, dx = F.$$

Hence if φ is rank-one convex

$$\varphi(F) \leq \int_{\mathbf{M}} \varphi(A)\, d\nu(A) = \int_{\Omega} \varphi(\nabla u)\, dx$$

and φ is, by definition, quasiconvex.

The proof of the theorem is not hard once is placed in the proper functional-analytical context. We only give the different ingredients, and refer the reader to [33]. On the one hand, we have the representation of the rank-one convexification obtained by B. Dacorogna in [11] (which led him to introduce the concept of (H_n) condition): if φ is continuous then its rank-one convexification, $\mathrm{R}\varphi$, is given by

$$\mathrm{R}\varphi(F) = \inf\left\{\sum_{i=1}^{n} \lambda_i \varphi(A_i) : \{(\lambda_i, A_i)\}_{1\leq i\leq n} \text{ verifies the } (H_n) \text{ condition and } F = \sum \lambda_i A_i \text{ for some } n\right\}.$$

A principal fact here is to have the uniform bound on the A_i's in the above infimum. This can be achieved by noticing the following fact which is almost an observation.

PROPOSITION. *Let $\varphi : \mathbf{M} \to \mathbf{R}$ be a continuous function such that $\varphi^{**} = \varphi$ off some closed, convex set B. Then if $F \in B$,*

$$R\varphi(F) = \inf\left\{\sum_{i=1}^{n} \lambda_i \varphi(A_i) : \{(\lambda_i, A_i)\}_{1\leq i\leq n} \textit{ verifies the } (H_n) \textit{ condition, } F = \sum \lambda_i A_i \textit{ for some } n \textit{ and } A_i \in B, \forall i\right\}.$$

φ^{**} denotes the convexification of φ, but the condition $\varphi^{**} = \varphi$ can be replaced by $\mathrm{R}\varphi = \varphi$.

All of this together with the Hahn-Banach theorem yields the result, although one has to proceed in two steps, considering first a special class of rank-one convex functions that take on the value $+\infty$ outside some compact set. The idea of the proof is the same for quasiconvexity and Young measures though a fact like the one in the proposition for quasiconvexity is missing, and some other technique is required to overcome this difficulty. We don't mean, though, that the proposition is not true for quasiconvexity. We simply don't know.

5. The three matrices case. The main fact we want to talk about in this final section is the case $\mathcal{K} = \{A_1, A_2, A_3\}$ where the A_i's are not pairwise rank-one related, and show that then there cannot be a laminate supported in $\mathcal{K}$ except δ_{A_i}. To put it in another way, if f is a continuous function with some suitable condition at infinity (for example if $f^{**} = f$ off some compact set) and $f^{-1}(0) = \mathcal{K}$, then $\mathrm{R}f^{-1}(0)$ is again $\mathcal{K}$ or $\mathrm{R}f$ is strictly positive in $\mathrm{co}(\mathcal{K}) - \mathcal{K}$.

The case $A_i - A_j$ =rank-one matrix can be easily handled. The diagrams would be

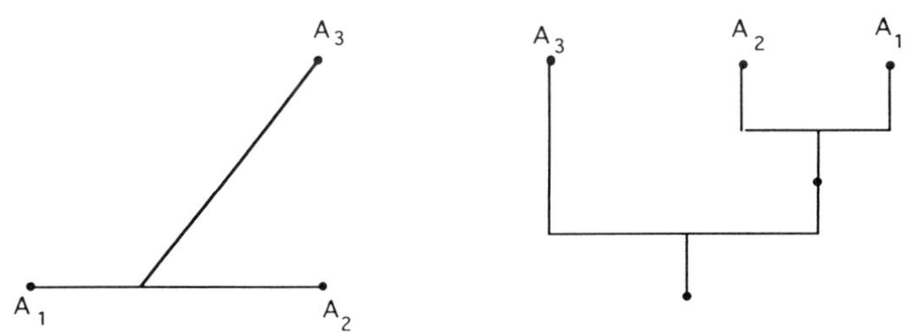

Figure 9

when $A_1 - A_2$ is rank-one and $A_3 - \lambda A_1 - (1-\lambda)A_2$ is also a rank-one matrix for some $\lambda \in (0,1)$.

We need to define the idea of quasiconvexification, rank-one convexification and even polyconvexification of sets. What we mean by each one of these convexifications is the following. Given a set of matrices $\mathcal{K}$, $\mathrm{R}\mathcal{K}$ (or $\mathrm{Q}\mathcal{K}$ or $\mathrm{P}\mathcal{K}$) denotes the set of matrices F, such that there is a laminate (or Young measure or polyconvex measure, respectively) with support in $\mathcal{K}$ and underlying deformation F. A polyconvex measure is a probability measure ν for which

$$\varphi(F) \leq \int_{\mathcal{K}} \varphi(A)\, d\nu(A)$$

for every polyconvex, continuous function. Or equivalently

$$\mathrm{M}(F) = \int_{\mathcal{K}} \mathrm{M}(A)\, d\nu(A)$$

for every minor M.

Obviously,

$$\mathrm{R}\mathcal{K} \subset \mathrm{Q}\mathcal{K} \subset \mathrm{P}\mathcal{K} \subset \mathrm{co}(\mathcal{K}).$$

It is not a simple matter to find R$\mathcal{K}$ or Q$\mathcal{K}$ even in "easy" cases. When $\mathcal{K}$ is a set of three matrices $\{A_1, A_2, A_3\}$ which are not pairwise rank-one related, the way to prove that no non-trivial laminate exists is by showing that R$\mathcal{K} = \mathcal{K}$. But actually, it is also true for some particular examples that R$\mathcal{K}$ is strictly contained in P$\mathcal{K}$ and we don't know in these cases what Q$\mathcal{K}$ is. The special structure of R$\mathcal{K}$ enables us to conclude that R$\mathcal{K} = \mathcal{K}$, for this very particular $\mathcal{K}$.

Several results on the structure of R$\mathcal{K}$ for more general situations are available in [33]. For the sake of simplicity, we are going to keep the requirements to a minimum and deal with the particular case we are treating. The crucial fact is the following:

THEOREM. *If $\mathcal{K} = \{A_1, A_2, A_3\}$ and A_1 is not rank-one related to anything in co($\mathcal{K}$) then R$\mathcal{K}$ $\subset$co($\{A_2, A_3\}$) $\cup \{A_1\}$.*

The idea is that in order to have non-trivial laminates in which A_1 participates, we need rank-one connections with the convex hull of $\mathcal{K}$, if not with $\mathcal{K}$ itself (as in the example with four matrices in section 3). As an intermediate step in the proof, we search for P$\mathcal{K}$. This is rather elementary for this $\mathcal{K}$ (restricting attention to dimension 2) since the only requirement is the linearity of det. The worst possible case (shown in Figure 10) is P$\mathcal{K} = \gamma \cup \{A_1\}$ where γ is a continuous curve through co($\mathcal{K}$) joining A_2 and A_3, and $\gamma \cap [A_2, A_3] = \{A_2, A_3\}$. In some other instances, according to how many rank-one lines lie in the plane spanned by the A_i's, we directly have P$\mathcal{K} = \mathcal{K}$.

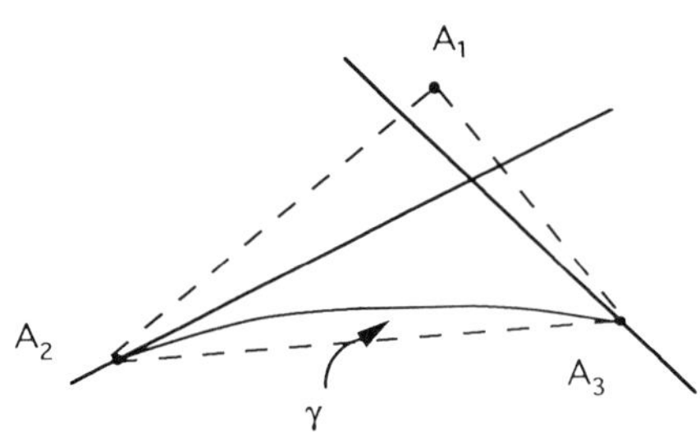

Figure 10

When $A_i - A_j$ is not rank-one for any combination of indices then it is elementary to realize that at least one of the A_i's (say A_3) is not rank-one to co($\mathcal{K}$), so that by the theorem R$\mathcal{K}$ $\subset$co($\{A_2, A_3\}$) $\cup \{A_1\}$. And since A_2 is not rank-one related to A_3 either, then R$\mathcal{K} = \mathcal{K}$. Consequently, the only laminates are the delta measures δ_{A_i}, as desired.

At this point we don't know whether there might be genuine Young measures supported in three matrices not pairwise rank-one related, but if there is a sequence generating such a parametrized measure it has to be so chaotic that we would have a very hard time discovering any microstructure or pattern in it. Very recently, V. Sverak has apparently found a proof that such an object can not exist at least in dimension 2.

REFERENCES

[1] J. M. Ball and R. James, *Fine phase mixtures as minimizers of energy,*, Arch. Rat. Mech. Anal., 100(1987), pp. 15–52..

[2] J. M. Ball and R. James, *Proposed experimental tests of a theory of fine microstructure and the two well problem.*

[3] K. Battacharya, *Self accomodation in martensite.*

[4] K. Battacharya, *Wedge-like microstructure in martensite.*

[5] M. Chipot and C. Collins, *Numerical approximation in variational problems with potential wells,.*

[6] M. Chipot and D. Kinderlehrer, *Equilibrium configurations of crystals*, Arch. Rat. Mech. Anal., 103 (1988), pp. 237–277.

[7] C. Collins and M. Luskin, *The computation of the austenitic-martensitic phase transition*, PDE's and continuum models of phase transitions, Lecture Notes in Physics, 344(1989), (Rascle, M., Serre, D., and Slemrod, M., eds) Springer, pp. 34–50.

[8] C. Collins and M. Luskin, *Numerical modelling of the microstructure of crystals with symmetry-related variants*, in *Proc. ARO US-Japan Workshop on Smart/Intelligent Materials and Systems, Technomic.*

[9] C. Collins and M. Luskin, *Optimal order error estimates for the finite element approximation of the solution of a nonconvex variational problem.*

[10] C. Collins, D. Kinderlehrer and M. Luskin, *Numerical approximation of the solution of a variational problem with a double well potential*, SIAM J. Numer. Anal. (to appear).

[11] B. Dacorogna, *Remarques sur les notions de polyconvexité, quasi-convexité et convexité de rang 1*, J. Math. Pures Appl., 64 (1985), pp. 403–438.

[12] B. Dacorogna, *Direct methods in the Calculus of Variations*, Springer, 1989.

[13] J. L. Ericksen, *On the symmetry of deformable crystals*, Arch. Rat. Mech. Anal., 72 (1979), pp. 1–13.

[14] J. L. Ericksen, *Some phase transitions in crystals*, Arch. Rat. Mech. Anal., 73 (1980), pp. 99–124.

[15] J. L. Ericksen, *Changes in symmetry in elastic crystals*, in *IUTAM Symp. Finite Elasticity (Carlson, D. E. and Shield R. T., eds) M. Nijhoff*, (1981), pp. 167–177.

[16] J. L. Ericksen, *Ill posed problems in thermoelasticity theory,*, in *Systems of Nonlinear Partial Differential Equations (Ball, J., ed)*, D. Reidel, 1983, pp. 71–95.

[17] J. L. Ericksen, *The Cauchy and Born hypotheses for crystals*, in *Phase Transformations and Material Instabilities in Solids, (Gurtin, M., ed)*, Academic Press, 1984, pp. 61–78.

[18] J. L. Ericksen, *Constitutive theory for some constrained elastic crystals*, Int. J. Solids Structures, 22 (1986), pp. 951–964.

[19] J. L. Ericksen, *Stable equilibrium configurations of elastic crystals*, Arch. Rat. Mech. Anal., 94 (1986), pp. 1–14.

[20] J. L. Ericksen, *Twinning of crystals I*, in *Metastability and Incompletely Posed Problems, IMA Vol. Math. Appl. 3, (Antman, S., Ericksen, J. L., Kinderlehrer, D., and Müller, I., eds)*, Springer, 1987, pp. 77–96.

[21] N. B. Firoozye, *Optimal Translations and Relaxations of some Multiwell Energies*, PhD Thesis, Courant Institute, New York University.

[22] N. B. Firoozye, *Optimal Use of the Translation Method and Relaxations of Variational Problems.*

[23] R. D. James, *Microstructure and weak convergence*, in *Proc. Symp. Material Instabilities in Continuum Mechanics, Heriot-Watt, (Ball, J. M., ed)*, Oxford, 1988, pp. 175–196.

[24] R. D. James and D. Kinderlehrer, *Theory of diffusionless phase transitions*, PDE's and continuum models of phase transitions, Lecture Notes in Physics, 344 (Rascle, M., Serre, D., and Slemrod, M., eds) (1989), pp. 51–84, Springer.

[25] D. KINDERLEHRER, *Remarks about the equilibrium configurations of crystals*, in *Proc. Symp. Material instabilities in continuum mechanics, Heriot-Watt (Ball, J. M., ed)*, Oxford, 1988, pp. 217–242.

[26] D. KINDERLEHRER AND P. PEDREGAL, *Characterizations of gradient Young measures*, Arch. Rat. Mech. Anal. (to appear).

[27] D. KINDERLEHRER AND P. PEDREGAL, *Remarks about Young measures supported on two wells.*

[28] D. KINDERLEHRER AND P. PEDREGAL, *Weak convergence of integrands and the Young measure representation.*

[29] R. V. KOHN, *The relationship between linear and non-linear variational models of coherent phase transitions*, in *Proc. of Seventh Army Conf. on Appl. Math. and Computing, West Point, June 1989.*

[30] R. V. KOHN, *The relaxation of a Double-Well Energy.*

[31] J. MATOS, *The absence of fine microstructure in* α-β quartz.

[32] J. MATOS, *PhD Thesis*, University of Minnesota.

[33] P. PEDREGAL, *Laminates and Microstructure.*

[34] P. PEDREGAL, *PhD Thesis*, University of Minnesota.

[35] A. C. PIPKIN, *Elastic materials with two preferred states*, Preprint, Aug. 1989.

MULTIPLY TWINNED PHASES AND MICROSTRUCTURES IN Ni-Al : A TRANSMISSION ELECTRON MICROSCOPY STUDY

D. SCHRYVERS*

Introduction. In today's understanding of the behaviour of common as well as new materials the knowledge at the atomic scale of microstructures and phase transitions is crucial. One technique that provides information at this level and which is not averaged over bulk size objects is transmission electron microscopy (TEM). In the direct imaging mode contemporary instruments can provide lattice images of a wide range of materials which enables one to investigate the atomic structure of the matrix as well as single defects at or below the unit cell dimensions; this mode is usually referred to as high resolution electron microscopy (HREM). In the electron diffraction (ED) mode the full symmetry at any given region down to a few nanometers in size can be found. Moreover, phase transitions can be followed in situ under, e.g., temperature or stress gradient conditions in dedicated specimen holders. These features, together with the possibilities for local chemical analysis by energy dispersive X-ray (EDX) analysis or electron energy loss spectroscopy (EELS) make this experimental technique extremely valuable in the investigation of phase transitions and the involved microstructures and phases.

In the present contribution the results of a typical TEM study are presented in some detail. It will become clear that some of the observed features can possibly be explained in terms of the nonlinear elasticity calculations of the energies of phase mixtures, although in some cases a competition with atomic scale forces can be expected. Other examples of TEM studies that exhibit microstructures resembling the outcomes of such calculations can, e.g., be found in reference [1].

The austenite-martensite transformation in Ni-rich Ni-Al alloys has been the subject of numerous experimental studies for over more than 30 years. The first report of a cubic to tetragonal transition dates back to Guseva & Makarov in 1951 [2]. Later Guard & Turkalo noted a different fracture behaviour between these two phases, the tetragonal phase being less brittle than the cubic one [3], while Maxwell & Grala first used the term "martensite-like" to describe the morphology observed in some quenched samples [4]. From then on a number of investigations focused on the microstructure of the tetragonal phase, now fully recognized as a martensite. Ball first suggested the existence of twinned plates [5] while Enami, Nenno & Shimizu showed internal twinning in these plates [6]. The morphology and crystallography of the twinning was discussed in detail by Chakravorty & Wayman [7]. Later different authors observed precursor effects of the transition in the austenite phase [8–10]. By now two different martensite structures are recognized: a 3R fct structure and a long period $(5\bar{2})$ microtwinned stacking of the former [11]. Although the shears and shuffles involved are thoroughly studied on the basis of phenomenological

*University of Antwerp, RUCA, Electron Microscopy for Materials Science, Groenenborgerlaan 171, B–2020 ANTWERPEN, Belgium

crystallographic theory, it is clear that the direct observation of the transition and microstructures in the EM will be of great value for understanding the local atom displacements and thus the process of the transition itself.

Most of the above mentioned investigations used conventional transmission electron microscopy to study the crystallography of this transformation, yielding results on habit plane orientation, twin morphology, microtwin arrangements etc. Also precursor effects typical for β-phase alloys have been observed and thoroughly documented. In the present work we try to obtain new information on the actual transition path by especially using high resolution electron microscopy.

The martensitic transformation in Ni-rich Ni-Al and its HREM image characteristics. In Ni-rich Ni-Al the high temperature parent or austenite phase ordered with CsCl type structure based on a bcc lattice can transform martensitically to a low temperature martensite phase with a CuAuI type structure based on a fct lattice; the crystallographic relation between both lattices is schematically depicted in Figure 1. The tetragonality of the martensite structure is dictated by its ordered character giving a layering of alternating Ni and Al planes in the $[001]_{fct}$ direction. Since the excess Ni atoms in Ni-rich Ni-Al are accommodated by substitution on the Al $(001)_{fct}$ planes, the composition of the material can be expected to have direct influence on the tetragonality of the martensite phase.

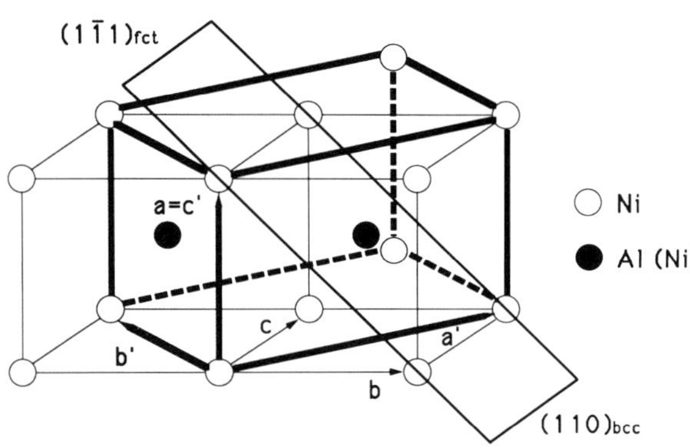

Figure 1: Crystallographic relation between the CsCl (fine) and CuAuI (bold type unit cells. The $(110)_{bcc}$ plane becoming the $(1\bar{1}1)_{fct}$ close packed plane is indicated.

The transformation is commonly described by a Bain type distortion during which one of the bcc cube axes is elongated while the other two are evenly shortened. This way four of the six bcc $\{110\}$ planes are distorted to become close packed fct $\{1\bar{1}1\}$ planes (see also Figure 1) and the amount of tetragonality is a function of the distortion. The amount of distortion will thus also be dictated by the composition.

From the $[001]_{fct}$ projections in Figures 2a and b it is clear that this distortion immediately implies a proper ABC stacking of close packed $(1\bar{1}1)_{fct}$ planes for the product phase. In order to fully describe the transition in this scheme one only needs to add a rigid body rotation which will depend on the actual choice of habit plane. In this way $3 \times 4 = 12$ variants of the martensite can evolve from the cubic parent structure. When microtwinning occurs in a given martensite plate both twin variants originate from the elongation along different cube axes of the parent phase; this is shown in Figure 2c.

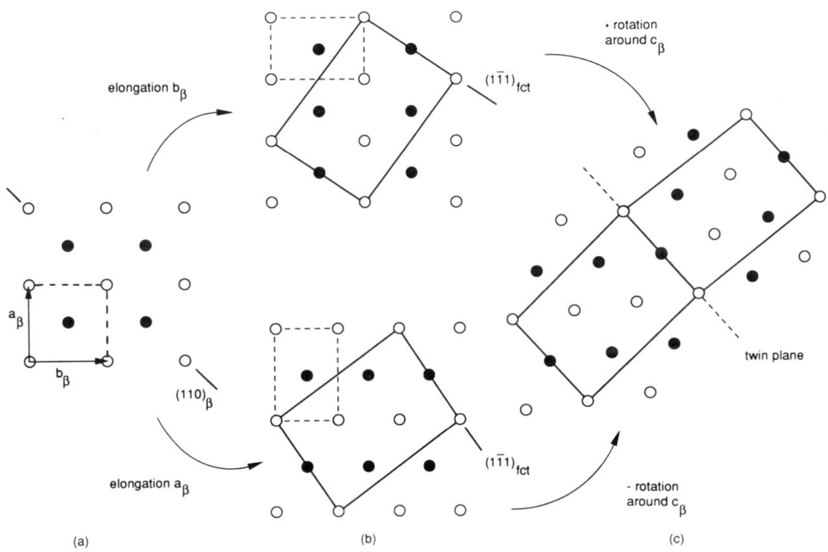

Figure 2: a) $[001]_{bcc}$ projection of the β austenite structure and b) $[110]_{fct}$ projections of two possible variants of the 3R martensite as obtained by elongation of the a or b axes of the bcc structure; o = Ni, • = Al(Ni). In both cases the same $(1\bar{1}1)$ plane is chosen as basal plane. c) Combination of the former variants into twin related crystals by rigid body rotations.

For an easy understanding of typical HREM images of these type of binary alloys the atomic structures involved are projected along the viewing direction in the instrument; such directions usually have low crystallographic indices (i.e., high symmetry) such as $[001]_{bcc}$ or $[110]_{fct}$ as in Figure 2. One thus obtains a two dimensional configuration, sometimes with higher symmetry than the corresponding plane of the crystal. In this pattern different sublattices can be recognized, depending on the structure as well as the chosen orientation. When the NiAl cubic austenite structure is viewed along the [001] direction, one finds a centered square pattern of pure Ni and mixed Al(Ni) columns (due to the non-stoichiometry) as shown in

Figure 2a. Due to the difference in scattering behaviour between Ni and Al atoms, these two types of columns can be visualized separately in a HREM instrument by choosing the proper imaging conditions. As a result one can expect a simple square pattern of white dots with edges parallel with [100] and [010]; depending on the actual imaging conditions the white dots will correspond with the Ni or Al subpattern, or possibly with the empty columns in between. However, since most discussions only involve information on the relative positions of these white dots the latter ambiguity will not affect the conclusions.

It follows that when a transition can be described by simple uniform distortions, the resulting structure will also yield a simple white dot image in a HREM instrument, possibly with the need for a slight reorientation of the crystal, but showing different angles and ratios. Indeed, when the fct martensite structure is viewed along its [110] direction, which will be nearly parallel with a cube direction of the parent structure, the white dot configuration will reveal the geometrical pattern of one of the sublattices in Figure 2b; here the bcc square has become a rectangle, as indicated by dashes.

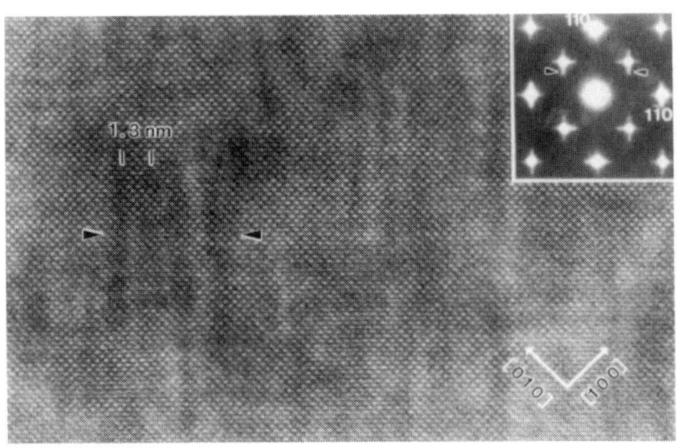

Figure 3: $[001]_{bcc}$ HREM image revealing $\{110\}\langle 1\bar{1}0\rangle$ micro modulations in different domains. The diffuse $\langle 110\rangle^*$ streaks in the diffraction pattern (inset) are the reciprocal analogs of these distortions.

The precursor state. Since the HREM images have to be taken at room temperature we have chosen a $Ni_{62.5}Al_{37.5}$ sample with M_s below room temperature in order to study the austenite phase. Here the quenched material is in the precursor state and the expected HREM image is slightly distorted. Indeed, although the square white dot pattern is still visible, numerous contrast variations parallel with (110) and $(1\bar{1}0)$ planes are observed, as seen from Figure 3. From some regions, indicated by arrows in the image, it can be inferred that these variations correspond to a modulation with a wavelength of approximately 1.3nm and a correlation length of about 4nm. From electron and optical diffraction [12] it can be concluded that these modulations are the image analogues of the $\langle 110\rangle^*$ diffuse streaks at Bragg reflections known to exist in precursor states of β-phase alloys [13].

The present modulations are, however, of a much finer scale than the conventional tweed striations [14] and are thus called micro-tweed or micro modulations. From the extinction conditions it is concluded that these modulations are of the transverse displacement type which is noted as $\{110\}\langle 1\bar{1}0\rangle$, indicating a $\{110\}$ plane being displaced in the $\langle 1\bar{1}0\rangle$ direction. The correspondence with the diffraction phenomenon mentioned above suggests that these modulations will also exist along the four remaining $\{011\}$ planes in the bcc structure. Due to the specific [001] orientation of the image one can, however, not expect to see those. From images such as the one shown in Figure 3 we conclude that the precursor state in Ni-Al consists of a contiguous patchwork of domains of a few nm in size, each accommodating a uniform $\langle 1\bar{1}0\rangle$ shear superimposed with a micro modulation with a 1.3nm wavelength and wave-vector in the $\langle 110\rangle$ direction perpendicular to the uniform shear. A suggestion for the actual atomic displacements inside the domains was recently presented on the basis of detailed image calculations [15].

At present this patchwork is attributed to the existence of numerous point-like defects inducing a stress field with a tetragonal component in the surrounding matrix. The matrix responds to this stress field following its soft modes, known to be $\{110\}\langle 1\bar{1}0\rangle$ with an extra dip at $q = 0.16\langle 1\bar{1}0\rangle^*$, the latter corresponding with the wavelength of 1.3nm [16].

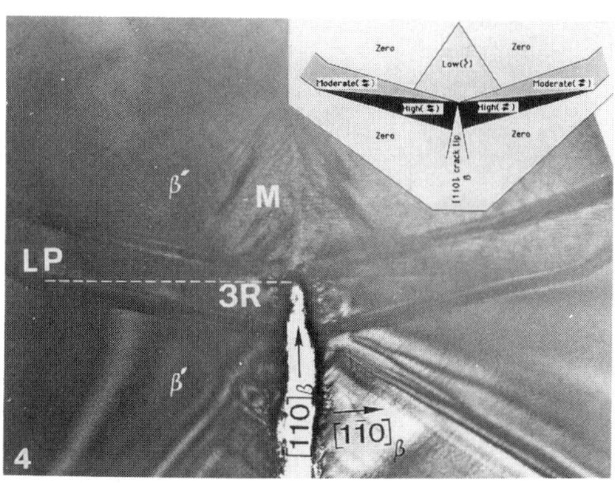

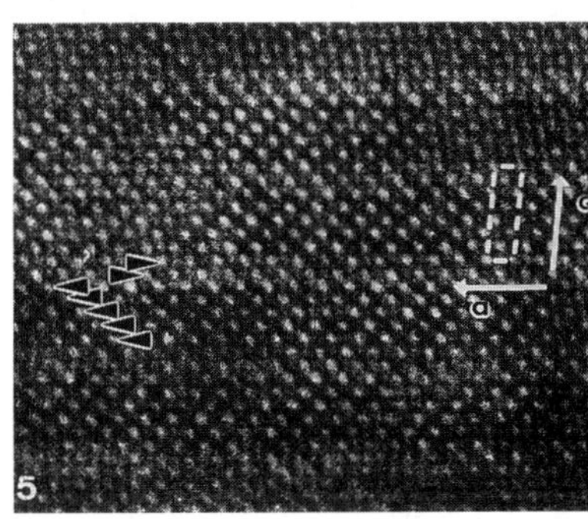

Figure 4: Microcrack in a thin foil edge with indications of different structures in the stress region around the tip of the crack. The inset gives a possible distribution of this stress.

Figure 5: HREM image of the 7M $(5\bar{2})$ stacking, as indicated by arrows.

The transition. As the material is cooled through the M_s temperature the martensite is invariably seen to nucleate at lattice defects such as dislocations or grain boundaries having significantly higher strain amplitudes than the point defects existing inside the uniform matrix. In samples thinned for EM observation the [001] foils often reveal cracks along the soft $\langle 110\rangle$ direction perpendicular to the foil edge. In some cases the stress field at the tip of such a crack is sufficient to induce the martensite transition at a temperature above the M_s for unstressed material. An example hereof is seen in the micrograph of Figure 4. Away from the tip the expected precursor contrast exists (β') while two wing shaped plates extend from the tip into the matrix. Inside these plates two sharply separated regions can be distinguished, one labelled 3R and having the expected martensite structure, the other being a heavily faulted long period (LP) version of 3R. In a similar sample we were able to obtain HREM images of the faulted region revealing the 7M microtwinned structure with some stacking faults; an example of the perfect $(5\bar{2})$ stacking is shown in Figure 5. These results indicate that the twin sequence can be influenced by the actual distribution of stress in a given region [16].

The martensite microstructure

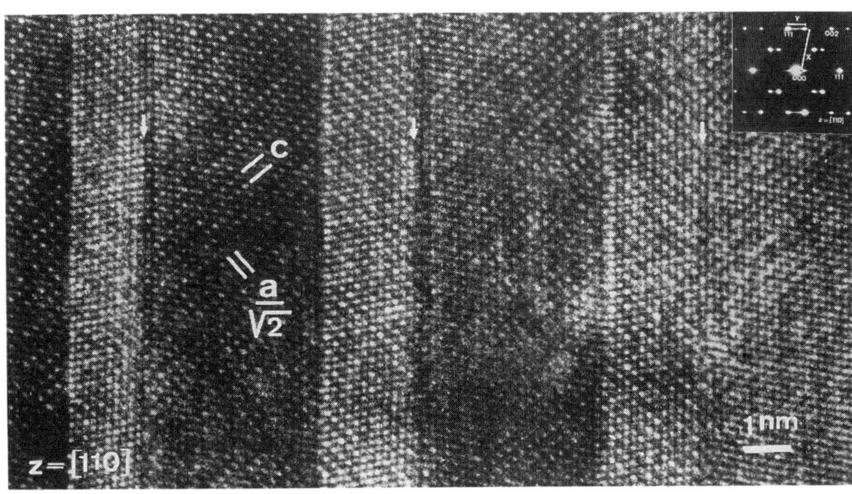

Figure 6: HREM image of a microtwinned 3R martensite with fct lattice dimensions measurable from this projection. The inset shows the corresponding ED pattern.

In order to study the martensite microstructure at different compositions, samples from a single polycrystalline ingot were annealed at different temperatures in the decomposition region between the austenite and the $L1_2$ phase. As result we thus obtained austenite with 66.0, 65.0 or 63.5 at% Ni; due to the pure displacive character of the transition this composition remains unchanged by quenching into the martensite phase at room temperature. All of the martensite plates in these samples were heavily twinned. From ED and HREM pictures such as the ones shown in Figure 6 we measured the lattice parameters and the c/a ratio of the martensite.

Microtwin densities were obtained from low magnification TEM images. After a proper averaging it can be concluded that the microtwin density T_d decreases with increasing Ni content, while the c/a ratio of the fct lattice increases, as shown in the graph of Figure 7. A detailed description of the results from this set of experiments is presented in [17].

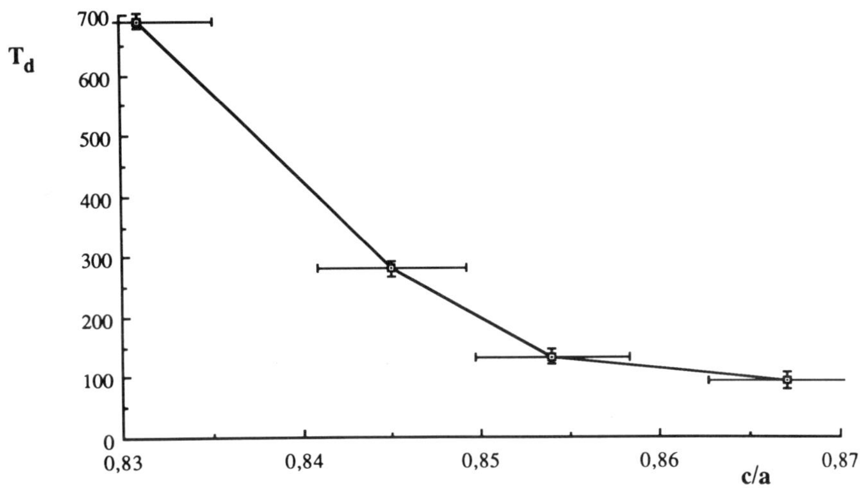

Figure 7: (c/a , T_d) graph showing the decrease of twin density with increasing c/a ratio of the fct martensite structure, i.e., with increasing Ni content.

Discussion. When one wants to consider the stability of a given atomic structure the essential internal parameters will consist of electronic contributions from the different constituents to the binding forces in the system. These forces will determine the Bravais lattice and its parameters as well as a possible ordered configuration. In some cases of order - disorder or diffusive transitions, i.e., where the basic lattice remains unchanged, one can calculate interaction parameters directly from the electronic configurations of the different atoms in the system [18]. The ratios of these parameters then determine the most stable ordered structure on the given lattice for a given composition without the need for an elaborate comparison of the total energies of all possible configurations [19].

In the case of displacive transitions, however, the change of basic lattice complicates such calculations in such a way that the energy of a limited number of simple structures has to be calculated separately and then compared [20]. Nevertheless, in certain systems or groups of systems a striking relation is observed between the valence electron per atom (e/a) ratio and the most stable stacking sequence of close packed planes. An explanation for this relation is then mostly sought in the size of the first Brillouin zone with respect to the Fermi surface [21]. A typical group exhibiting this type of behaviour consists of the AB_3 compounds of Ti or V with other transition elements, with $TiNi_3$ as well known example [22]. The sometimes complicated stackings of these phases are often denoted with their percentage of hexagonality, i.e., the number of hcp units in the entire unit cell chosen with a close packed plane as basal plane. The $(5\bar{2})$ stacking in Ni-Al could thus also be viewed in a similar scheme and having a hexagonality of 28.6%.

The above considerations on the energy of the structures involved do not take into acçount the dynamics of the transition. However, the electronic contributions will still solely determine the basic lattice and its parameters for an austenite phase and they will at least influence those of a martensite phase. For the β phase alloys it is clear that the interatomic forces, and thus the composition, will affect not only the lattice parameters but also the soft mode behaviour of the austenite phase which has a CsCl structure stable around $e/a = 3/2$. In the case of Ni-Al, the dip at $q=0.16\langle 110\rangle^*$ in the transverse acoustic branch of the phonon dispersion curve is most probably correlated with the observation of the micro modulation of 1.3nm in HREM [15]. This micro modulation can in turn be seen as a precursor to the $(5\bar{2})$ stacking of the martensite phase in $Ni_{62.5}Al_{37.5}$. In this example it thus seems likely to consider the 7M structure, be it a micro-twinned 3R structure, as a genuine phase with a small (meta)stability range between hcp and fcc determined by the electronic configuration of this compound. However, if the periodicity would be the only requirement, i.e., to get the proper Brillouin zone, one would expect a $(4\bar{3})$ stacking sequence instead of $(5\bar{2})$ [23]. From the proposed atomic configuration for the distorted domains in the precursor state [15] the $(5\bar{2})$ stacking can indeed be interpreted by the combination of a micro modulation shuffle with a wavelength of 7 $\{110\}$ planes and a uniform shear favouring one of both directions in this modulation. If this is the case the $(5\bar{2})$ stacking is the result of a competition between the strain energy and the wavelength imposed by the soft mode behaviour as dictated by the electronic contributions.

When the composition and with it the electronic configuration of the compound is changed the soft mode behaviour will also change and the dip might become less pronounced, resulting in a decreasing influence of the precursing micro modulation onto the stacking of close packed planes in the martensite phase. At this point the classical arguments of stress accommodation during the displacive transformation and inducing internal transformation twins or stacking faults will become more important. In Ni-Al the width of the microtwin variants rapidly increases to about $20 - 50$ close packed planes when the amount of Ni increases to about 66 at%. As the microtwinned structure is now considered as consisting of different variants of 3R combined in order to accommodate an internal stress configuration one should rather speak of a microstructure instead of a new metastable phase.

Especially for the latter compositions the calculations based on non-linear elasticity theory presented during this workshop can be of great interest. For example, the determination of the ratio λ of the width of the largest variant with the total width of both variants as a function of the parent and product lattice parameters could explain, e.g., the change in twin density [24]. Also, since the calculations on the minimizing sequences involve the energy of the internal twin planes [25] (with hcp configuration in the present case), ab initio calculations of the energy of hcp versus fcc units might yield a lower limit for the width of the smallest variant. The latter combined with λ then determines the stacking sequence with the lowest energy. However, for now the experimental information on the lattice parameters of parent and product phases at the transition is not sufficiently accurate. Indeed, as shown in [17], the measured (from EM micrographs) nor the calculated (from $a\beta$ and $a_m, c_m)\lambda$, with their respective accuracies, change over the investigated composition range. This is primarily due to the averaging procedure over a number of electron micrographs for a given treatment which is necessary since the observed microstructures do show the appreciable stacking disorder. The future aim will be to prepare samples with more perfect stackings over wide areas of the sample by more sophisticated thermal treatments.

Conclusion. On the basis of the results of an electron microscopic investigation of the martensitic transformation in Ni-Al we have tried to recapture some of the arguments on the stability of microtwinned structures. Depending on the particular situation, the intrinsic electronic contributions to the energy might be expected of greater or lesser importance than strain energy terms related with the transition path. As more data is gathered, experimental as well as theoretical, one might hope to be able to pin down the influence of both, essentially different, parameters on the structure of the martensite phases.

Acknowledgements. The author gratefully acknowledges the cooperation with Dr. L.E. Tanner. This research program was partially performed under the auspices of the U.S. Department of Energy by the LLNL under contract W–7405–ENG–48.

REFERENCES

[1] Verwerft, M., Van Tendeloo G., Van Landuyt J. and Amelinckx S., Ferroelectrics **97** (1989), 5.

[2] Guseva, L.N. and Makarov, E.S., Dokl. Akad. Nauk SSSR **77** (1951), 615 (abstr. transl. in Chem. Abstr. **47** (1951) 6211).

[3] Guard, R.W. and Turkalo, A.M., in *Mechanical Properties of Intermetallic Compounds*, J.H. Westbrook ed., 141 (John Wiley, NY) (1960).

[4] Maxwell, W.A. and Grala, E.M., NACA TN **3259** (1954), 39.

[5] Ball, A., Metal Sci. J **1** (1967), 47.

[6] Enami, K., Nenno, S. and Shimizu, K., Trans. Japan, Inst. Metals **14** (1973), 161.

[7] Chakravorty, S. and Wayman, C.M., Metall. Trans. **7A** (1976), 555, 569.

[8] Enami, K., Hasunuma, J., Nagasawa, A. and Nenno, S., Scripta Metall. **10** (1976), 879.

[9] Reynaud, F., Scripta Metall. **11** (1977), 765.

[10] Robertson, I.M. and Wayman, C.M., Metall. Trans **15A** (1984), 1353.

[11] Martynov, V.V., Enami, K., Khandros, L.G., Tkachenko, A.V. and Nenno, S., Scripta Metall. **17** (1983), 1167.

[12] Van Tendeloo, G. and Amelinckx, S., Scripta Metall. **20** (1986), 335.

[13] Delaey, L., Perkins, A.J. and Massalski, T.B., J. Mat. Sci. **7** (1972), 1197.

[14] Robertson, I.M. and Wayman, C.M., Phil. Mag. **48** (1983), 421, 443, 629.

[15] Schryvers, D. and Tanner, L.E., Ultramicroscopy **32** (1990),241.

[16] TANNER, L.E., SCHRYVERS, D. AND SHAPIRO, S.M., Mat. Sci. and Eng. **A 127** (1990), 205.

[17] SCHRYVERS, D., DE SAEGHER, B. AND VAN LANDUYT, J., Mat. Res. bull. **26** (1991), 57.

[18] BIEBER, A. AND GAUTIER, F., Sol. St. Comm. **38** (1981), 1219.

[19] KANAMORI, J. AND KAKEHASHI, Y., J. de Physique **C7** (1977), 274.

[20] TURCHI, P.E.A., Mat. Sci. and Eng. **A127** (1990), 145.

[21] SATO, H., TOTH, R.S. AND HONJO, G., J. Phys. Chem. Solids **28** (1967),137.

[22] SINHA, A.K., Trans. Metall. Soc. of AIME **245** (1969), 911.

[23] FUJIWARA, K., J. Phys. Soc. Japan **12** (1957), 7.

[24] BHATTACHARYA, K., these proceedings.

[25] BALL, these proceedings.

PHASE TRANSFORMATIONS IN β-PHASE SHAPE MEMORY ALLOYS STUDIED BY HRAEM, ALCHEMI AND X–RAY DIFFRACTION

KEN'ICHI SHIMIZU*

Abstract. Recent research results on the microstructures and phase transitions in some β-phase shape memory alloys are reviewed, which have been obtained by the author's research group by examining with high resolution analytical electron microscope (HRAEM), atom locations by channeling enhanced microanalysis (ALCHEMI) and X–ray diffraction. The HRAEM and ALCHEMI are first outlined, because the former is specially designed new one and enables us to obtain high resolution transmission electron microscope images, sharp electron diffraction patterns and non-contaminated energy dispersive X–ray spectra from very small areas less than 5 nm in diameter, and the latter is able to determine the atom location of a third element of ternary ordered alloys or compounds.

Then, bainitic transformations in Cu–29.1Zn–6.7A1 (at%) and Ag–45.0Cd (at%) shape memory alloys were examined by the HRAEM as to the microstructural, crystal structural and compositional changes. Ordering process and subsequent martensitic transformation in a Cu–Au–Zn shape memory alloy with the stoichiometric composition $CuAuZn_2$ have also been examined by the ALCHEMI and X–ray diffraction aiming at the atom configuration. The results obtained from the examinations are summarized, and brief discussion is made in relation to the mechanisms of phase transformations and shape memory characteristics.

1. OUTLINE OF HRAEM AND ALCHEMI METHODS

(1) HRAEM method. In this section, outline of the newly developed HRAEM will be briefly described about the performance capability. In the conventional analytical electron microscope (AEM) so far utilized, there have been two big problems. One is that sharp electron diffraction (ED) patterns were not able to be obtained because incident electron beams were strongly converged onto about 20 nm size areas of specimen surface. The other was that those areas were contaminated with a hydrocarbon material. Thus, because of the lack of sharpness in ED patterns and the contamination effect in EDX spectroscopy, the conventional AEM could not be used so effectively as might be desired when one was attempting the identification of materials with a complicated structure.

In order to solve those two problems, the areas to be analysed must be illuminated with less-converged electron beams and the inside of electron microscope column must be maintained in an ultrahigh vacuum. The best way to do so is through development of an AEM with a field emission (FE) electron gun. This is because the beam source is very small, namely 1 nm or less in size, and thus nm size areas on specimen surface can be illuminated by the electron beams without the need for much convergence. Moreover, the use of FE electron gun requires that the inside of electron gun is maintained in an ultrahigh vacuum less than 10^{-8} Pa. Thus, an FE electron gun was attempted to mount on a conventional AEM.

Figure 1 shows the system configuration of the newly developed HRAEM, Model H–600FE. Except for the FE electron gun and ultrahigh vacuum system, the total

*The Institute of Scientific and Industrial Research, Osaka University, 8–1, Mihogo–oka, Ibaraki, Osaka 567, Japan

system is almost the same as that of the conventional AEM, Model H-600. Since the beam source of the FE electron gun is very small, as mentioned before, the beam direction is affected very sensitively by external magnetic fields, such as a geomagnetism. So, the microscope columns of the FE electron gun, illumination lens system and specimen chamber were made of a permalloy, in order to shield from the external magnetic fields. By using this HRAEM, we can take transmission electron microscopy (TEM), scanning TEM (STEM) and scanning electron microscopy (SEM) images, energy dispersive X-ray (EDX) spectra and electron energy loss spectroscopy (EELS).

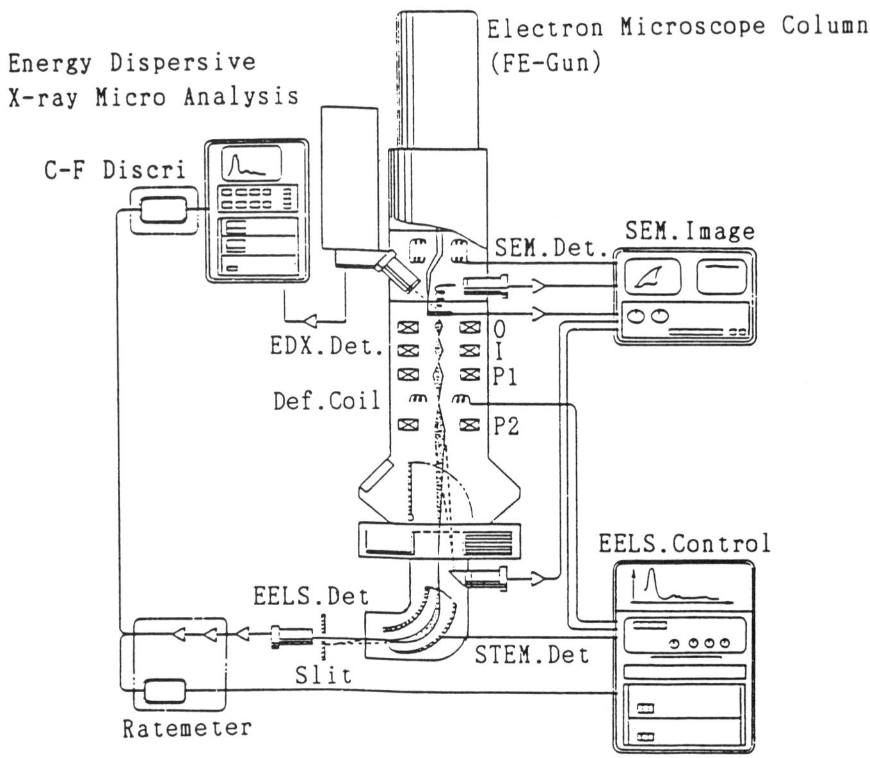

Figure 1. System configuration of Model H-600FE analytical electron microscope.

Table 1 shows the performance specifications of Model H-600FE HRAEM. The most important feature of the FE electron gun is high brightness. The brightness of a single crystal tungsten field emitter is 10^6 to 10^7 A/m^2, and is 500 times brighter than that of LaB_6 thermionic emitter. The next feature of the FE electron gun is a small point source of 1 nm or less, and this is far smaller than the typical one of 30 μm for the ordinary hairpin type thermionic cathod. The third feature is a small energy spread of 0.2 eV, and this is one tenth of the normal thermionic source. Other specifications are shown in the table. The above three features as well as other specifications bring about many improvements in AEM studies, such as higher contrast in imaging and high resolution in ED, EDX and EELS studies of nm size areas. According to the lens system adopted in the developed HRAEM, beam size on specimen surface is approximately 20 nm in the TEM mode, and 5 nm or less in micro ED or EDX mode, as shown in the table. Thus, high resolution TEM images, sharp ED patterns and non-contaminated EDX and EELS spectra are able to be taken from very small areas, and then the HRAEM enables us to make the characterization of materials with a complicated structure possible more precisely than the conventional AEM. Details of the HRAEM are given in Ref. (1), and readers should refer to it.

Table 1. Performance specifications of the Model H–600FE analytical electron microscope used in the present study.

Accelerating voltage	Analytical TEM (100 kV)	High-resolution TEM (100 kV)
Resolution		
TEM lattice/point	0.14 nm/0.45 nm	0.14 nm/0.30 nm
SEM	< 1.0 nm	
STEM	< 1.0 nm	
Specimen tilt angle	$\pm 60°(\pm 25^{oa})$	
Magnification		
TEM		
LM	$100\_ \times 500$	$100\_ \times 400$
HM	$1000\_ \times 300,000$	$5000 - \times 800,000$
SA	$2000\_ \times 100,000$	$2000\_ \times 200,000$
SEM	$100\_ \times 500,000$	
STEM	$100\_ \times 500,000$	
ED camera length		
SA	0.2–1.6 m	0.1–0.8 m
SHD	10–100 m	10–100 m
SAED		
Spot size	< 5 nm θ	20 nm θ^b
Resolution	$< 10^{-4}$	$< 10^{-4b}$
EELS		
Energy resolution	< 1 eV	< 1 eV[c]
X-ray	High-angle (68°) take-off detector	

[a] EDX mode.
[b] TEM SAED mode.
[c] TEM EELS mode.

(2) ALCHEMI method

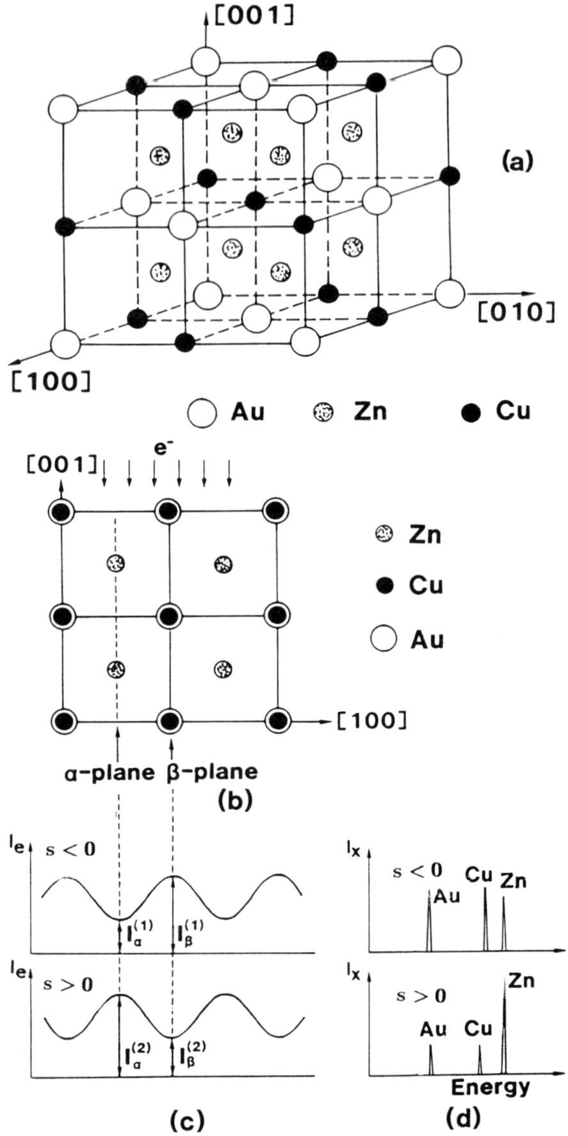

Figure 2. Atom configuration in the unit cell of the $L2_1$ type ordered $CuAuZn_2$ crystal, (a), atom configuration projected onto the (010) plane of the $CuAuZn_2$ crystal, (b), channeling enhancement of the electron beam intensity averaged over crystal thickness for $s > 0$ and $s < 0$, (c), and characteristic X-ray intensities from Cu, Au and Zn atoms for $s > 0$ and $s < 0$, (d).

Next, the ALCHEMI method will be explained referring to $L2_1$ type ordered $CuAuZn_2$ alloy, which will be described later in more detail. Figure 2(a) shows an atomic configuration of the unit cell of the $L2_1$ ordered alloy, and (b) does the projected one onto the (010) plane. There are seen two kinds of (200) planes, one consisting of only Zn atoms and the other doing of both Cu and Au atoms. The two kinds of (200) planes will be called α-plane and β-plane, respectively. Corresponding to the existence of such two kinds of (200) planes, 200 superlattice reflection appears in $\langle 100 \rangle$ zone ED patterns. In such a two-beam diffraction condition that the 200 reflection is only excited, incident beams form standing waves, and a channeling effect is caused resulting in an enhancement of incident beams along the α-plane or β-plane depending on diffractions. As shown in (c), the intensity of incident beams averaged over crystal thickness is enhanced along the α-planes for a positive excitation error ($s > 0$), while it is along the β-planes for a negative excitation error ($s < 0$). Accordingly, the intensity of EDX spectra emitted from atoms on those planes should be changed by diffraction conditions, for example the intensities of Cu–K_α and Au–M_α spectra are stronger than that of Zn–K_α spectrum under the condition of negative s, and are weaker under the condition of positive s, respectively, as shown in (d).

As will be discussed later, if Zn atoms were replaced by Cu atoms, Zn–K_α spectrum would be stronger and weaker than Cu–K_α spectrum under the conditions of negative and positive s, respectively. Cu and Zn atom sites are generally very difficult to be distinguished by conventional X–ray diffraction, because their atomic scattering factors are very similar to each other. However, the ALCHEMI can clearly distinguish the Cu and Zn atom sites. Similarly, the ALCHEMI is very useful in determining atomic sites of the third or fourth alloying elements in binary or ternary superlattice alloys or intermetallic compounds.

2.BAINITIC TRANSFORMATION IN CU–ZN–A1 AND AG–CD ALLOYS

Recently, the bainitic transformations have again been paid a good deal of attention, because they play an important role in some newly developed functional materials [(2)]. Then, they have so far been extensively studied by using the conventional AEM. However, some discrepancy exists as to the compositional information of the bainite in Cu–Zn alloys [(3)] [(4)], and doubts have often been thrown upon the validity of the AEM studies. One of them arises from a spreading of electron beams in thin foil specimens, that is, EDX spectra come entirely from a bainite plate itself or both the plate and its adjacent matrix. Such an effect of beam spreading must be taken into consideration in AEM studies.

Then, the author's group has studied the bainitic transformations in Cu–Zn–A1 and Ag–Cd shape memory alloys by using the new HRAEM mentioned above. In the studies, ED patterns and EDX spectra have been taken from very thin bainite plates just after the formation.

(1) Cu–Zn–A1 alloy. The alloy used has the composition of Cu–29.1Zn–6.7A1 (at %), whose M_s temperature was 225 K after the homogenization and solution treatment at 1123 K for 1.8 ks and 300 s, respectively. Disks 3 mm in diameter were spark–cut from the homogenized and solution treated alloy, and aged at 473 K or 573 K for various periods ranging from 5.4 ks to 2.4 Ms to produce bainitic plates in the alloy. Other experimental procedures have been described elsewhere (5).

In the EDX spectroscopy, compositions of bainite plates and their adjacent matrices were quantitatively analysed by using a formula proposed by Cliff and Lorimer (6). k factor in the formula has been reported to vary with various experimental conditions. However, our interest is whether or not the composition of a bainite plate just after the formation is different from that of the adjacent matrix. So, particular attention was paid to the difference in Cu content between the two phases rather than the composition itself. Then a slight variation in k factor does not cause serious problem in the present experiment, because the analyses of the bainite and matrix were carried out under almost the same condition.

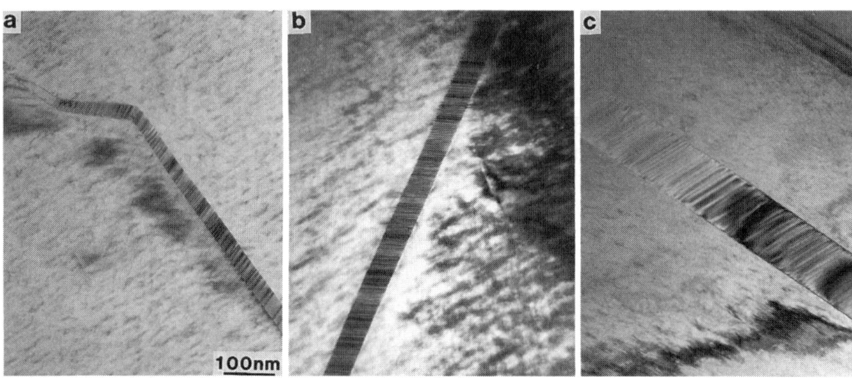

Figure 3. TEM images of thin bainite plates in a specimen heat-treated at 473 K for 7.2 ks, (a) and (b), and 28.8 ks, (c).

Figure 3 shows TEM images of thin bainite plates observed in specimens aged at 473 K, (a) and (b) being those aged for 7.2 ks and (c) being that aged for 28.8 ks. The bainite plates appear to correspond to an early stage of growth, and they have characteristic features of martensitic shear products, because parallel straight interfaces and a lot of internal stacking faults are seen. Figure 4 shows the TEM image of a bainite plate observed in a specimen aged at 473 K for 28.8 ks and the corresponding ED patterns. (b) was taken from the central bainite plate, and reflections can be indexed in terms of a M9R structure like those from the martensites formed at lower temperatures. Superlattice reflections are hardly seen in (b), but they can clearly be seen in other patterns under a systematic excitation condition. The similarity to martensite in crystal structure as well as microstructure suggests that the bainite is formed by a shear mechanism rather than by a diffusional process. (c) was taken from the adjacent matrix marked with M in (a), showing a

B2 type structure. Figure 5 is a high resolution TEM image of an interface between bainite plate and its adjacent matrix. Lattice fringes corresponding to the $(11\bar{4})$ planes of the bainite are connected to those of the $(0\bar{1}1)$ planes of the matrix at the interface, indicating that the bainite is coherent with the matrix at least up to about 50 nm thickness.

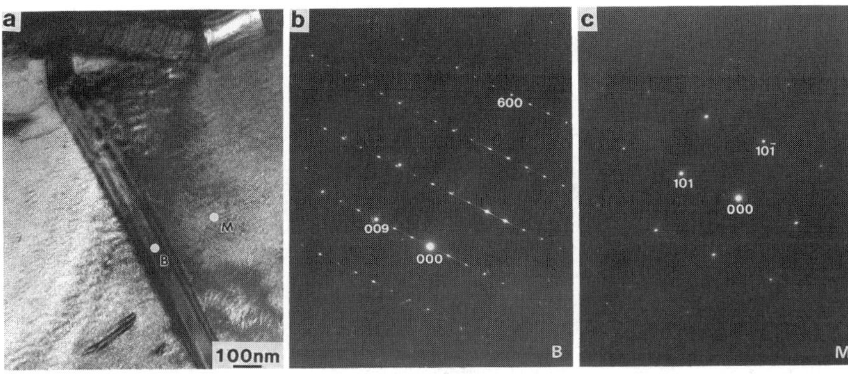

Figure 4. TEM image of bainite plates in a specimen heat-treated at 473 K for 28.8 ks, (a), and ED patterns taken from the central plate and its adjacent matrix, (b) and (c).

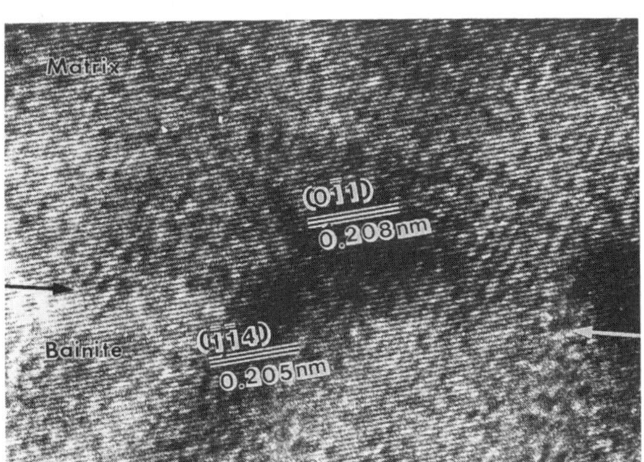

Figure 5. High resolution TEM image around an interface between bainite and matrix in a specimen heat-treated at 473 K for 10.8 ks.

Figure 6(a) is the TEM image of a bainite plate observed in a specimen aged at 473 K for 2.4 ks. Interfaces of the plate are not so straight as seen in Figure 3. Then, the coherence between the bainite plate and adjacent matrix seems to have already been lost in this stage of growth. (b) and (c) are ED patterns of the bainite and matrix regions marked with black spots, B and M. Superlattice reflections are no longer observed in (b), although the matrix has still B2 ordered structure, as superlattice reflections can be seen in (c). That is, the bainite has changed from an ordered M9R structure to a disordered N9R structure during the isothermal growth. (d) and (e) are EDX spectra taken from the bainite plate and the matrix, respectively, which will be explained later.

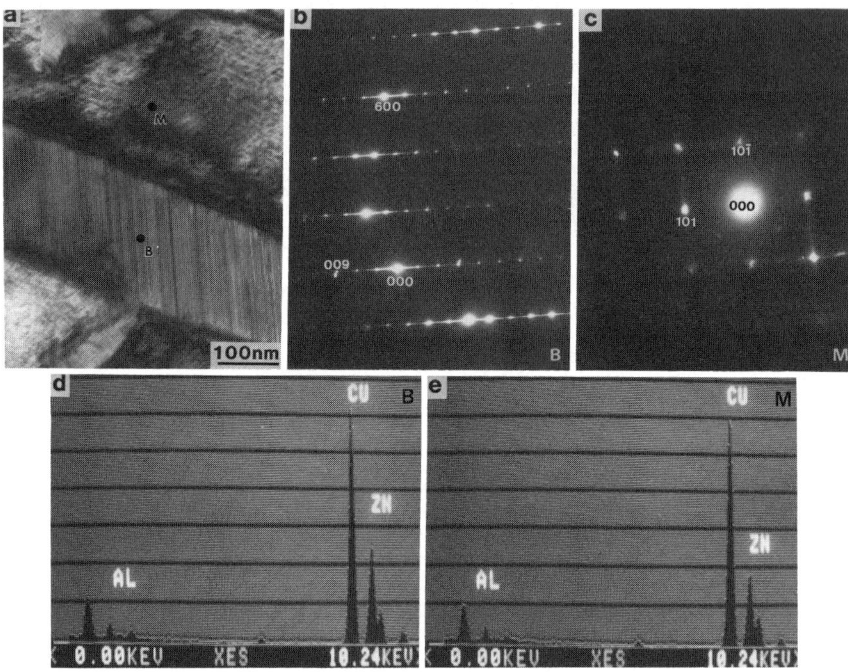

Figure 6. TEM image of a bainite plate in a specimen heat-trated at 473 K for 2.4 Ms, (a), ED patterns taken from the bainite and the matrix regions marked with black spots, (b) and (c), and EDX spectra taken from the black spot areas, (d) and (e).

Figure 7(a) is the TEM image of a bainite plate obtained by aging at 573 K for 604.8 ks. In this stage of growth, density of internal stacking faults in the bainites becomes much lower than that of the bainites previously shown. The upper side of interfaces of the bainite plate in (a) no longer forms a plane. Corresponding ED pattern (b) of the plate is indexed by a disordered fcc structure, which is equivalent to the equilibrium α phase as the final product during aging. (c) is an ED pattern corresponding to the adjacent matrix marked with M in (a). (d) and (e) are again EDX spectra of the bainite and the matrix, respectively, which will also be explained later in more detail.

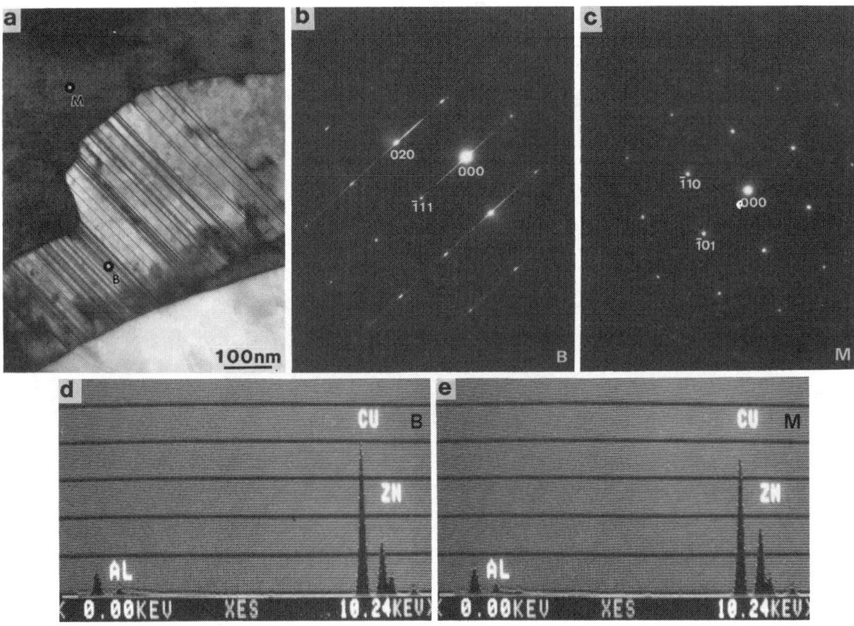

Figure 7. TEM image of a bainite plate in a specimen heat-trated at 573 K for 604.8 ks, (a), ED patterns taken from the bainite and the matrix regions marked with black circles, (b) and (c), and EDX spectra taken from the black circled areas, (d) and (e).

Figure 8 is an example of series of TEM image and EDX spectra. (a) is the TEM image of a bainite plate with about 30 nm thick, and then the bainite appears to have a M9R structure, which is the same as that of subzero cooled martensites, and to correspond to a very early stage of growth. (b) and (c) are the EDX spectra taken from the bainite and its adjacent matrix in (a), which have been taken from the areas marked with black spots of 10 nm diameter. As known from a comparison of the intensities of X–rays emitted from Cu and Zn atoms in (b) and (c), compositions of Cu and Zn atoms are already different between the bainite and adjacent matrix, although the bainite is very similar to subzero cooled martensites in microstructure and crystal structure. EDX spectra already shown in Figs. 6(d) and (e) were taken from a disordered N9R bainite plate and its adjacent matrix. These EDX spectra also show distinct difference in composition between the bainite and adjacent matrix. EDX spectra shown in Figs. 7(d) and (e) were taken from a disordered fcc bainite and adjacent matrix, which may correspond to a final state of bainite growth, and the bainite may be equivalent to the equilibrium α phase. It is then natural that those spectra show distinct difference in composition between the bainite and adjacent matrix. In this way, EDX spectra from bainite plates with various thickness and their adjacent matrices have been taken, and the difference in Cu content between them have been plotted as a function of plate thickness, because the thickness is considered to present more appropriately the period which has elapsed after the bainite formation, as shown in Figure 9. The figure indicates that the bainite plates are produced with a different composition from their matrices and grow keeping the composition. The most right hand side point was obtained from a bainite plate nearly corresponding to the equilibrium fcc phase. Then, roughly speaking, the bainite is considered to be formed with nearly the same composition as the equilibrium fcc α phase from the beginning of bainite formation.

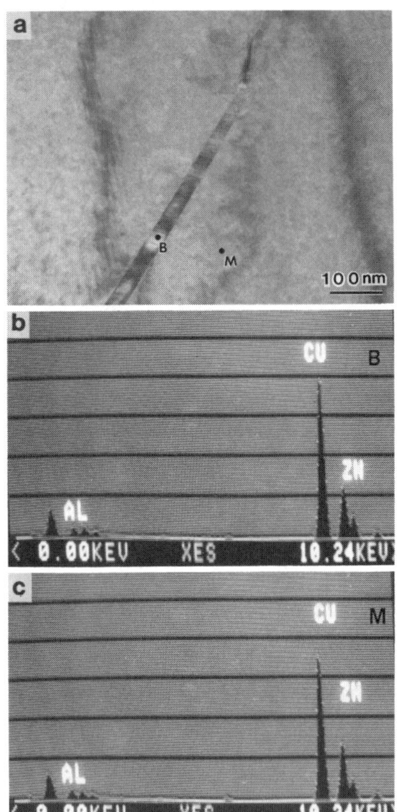

Figure 8. TEM image of a bainite plate in a specimen heat-trated at 473 K for 7.2 ks, (a), and EDX spectra taken from the bainite and the matrix regions marked with black spots, (b) and (c).

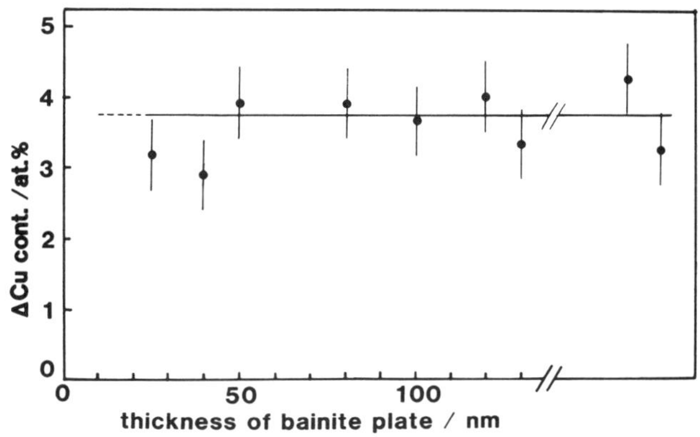

Figure 9. Difference in Cu content between bainites and as-quenched matrix, as a function of thickness of the bainites.

In this way, the composition of bainite is different from that of the adjacent matrix even at an early stage of growth, while the microstructure and crystal structure are almost the same as those of martensites. These results are apparently inconsistent with each other in view of the diffusionless nature of martensitic transformation. However, they become consistent if one considers that diffusion first occurs in local regions of the matrix prior to the bainite formation, causing some decrease of solute atoms in the regions, and that M_s temperature increases as a result and a martensitic transformation takes place in the local regions even at the temperature of 473 K higher than M_s.

(2) Ag–Cd alloy. The alloy used has the composition of Ag–45.0Cd (at%). 3 mm diameter and 0.3 mm thick sheets were cut from a 3 mm diameter ingot annealed at 823 K for 86.4 ks. The sheets were then betatized at 955 K for 300 s in air, and rapidly quenched into 260 K by compulsory dropping. M_s temperature of the quenched alloy was 218 K. The quenched sheets were then heat-treated for the bainitic transformation in a silicon oil bath at 433 K for various periods up to 1468.8 ks (17 days). The bainitic products thus obtained have been examined by TEM observation and EDX spectroscopy. Other experimental procedures should be referred to Ref. (7).

Figure 10 shows optical micrographs of the as–quenched matrix, (a), bainitic products formed by subsequent heat-treatment at 433 K for 3.6 ks, (b), and martensitic products formed upon subzero cooling of the as-quenched matrix, (c), in the Ag–Cd alloy. Microstructures of the bainites and the martensites in (b) and (c), respectively, are quite different from each other, being different from the case of the previous Cu–Zn–Al alloy. According to a previous study [8], the spear shaped martensites are γ_2' with a 2H type stacking order structure, and their internal substructures are twins.

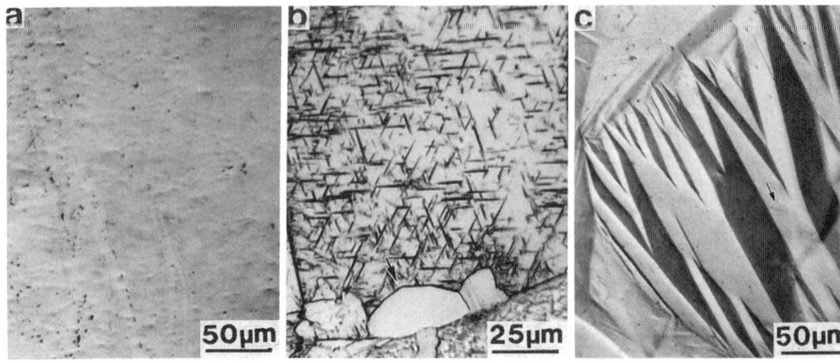

Figure 10. Optical micrographs of a Ag–45.0at%Cd alloy as quenched, (a), the bainites plate formed in the alloy subsequently heat-treated at 433 K for 3.6 ks, (b), and the γ_2' martensites produced in the as–quenched alloy upon subzero cooling to 183 K, (c).

Figure 11 shows the TEM image of a bainite plate with 50 nm thick in a specimen heat-treated at 433 K for 5.4 ks, and the corresponding ED patterns taken from the bainite plate and its adjacent matrix. Another similar series of TEM image and ED patterns taken of a bainite plate with 200 nm thick in the same specimen exhibited the same results as those in Figure 11. That is, the two series of TEM image and ED patterns showed that the bainite plates exhibited internal stacking fault fringes, and that their interfaces were parallel and nearly straight, being similar to those of 9R or 18R type martensites in other β-phase alloys. ED patterns (b) and (c) in Figure 11 have been taken from the black spot areas (5 nm diameter) in the bainite plate and its adjacent matrix, respectively, in TEM image (a). (b) shows that the bainite plate is of a 9R type long period stacking order structure, its c^* axis appearing on the pattern, and (c) does that the matrix is of a B2 type ordered structure, as known from the appearance of arrowed superlattice reflections. However, no superlattice reflections are observed in (b), and then the bainite is concluded to be disordered orthorhombic N9R structure, even though the martensite formed at lower temperatures was of a 2H type and the bainite in the previous Cu–Zn–A1 alloy was of an ordered monoclinic M9R structure.

Figure 12(a) shows the TEM image of a bainite plate whose thickness and stacking fault density increased and decreased, respectively, with the progress of growth. (b) and (c) are the corresponding ED patterns taken from the bainite and its adjacent matrix, marked with black spots B and P, respectively, in (a). (b) can be explained to be of a fcc structure, and (c) can be of a B2 phase. Therefore, the bainite plate is supposed to have changed from a disordered N9R to a disordered fcc structures, being different from the case of Cu–Zn–A1 alloys. The disordered fcc structure may probably correspond to that of the equilibrium α phase.

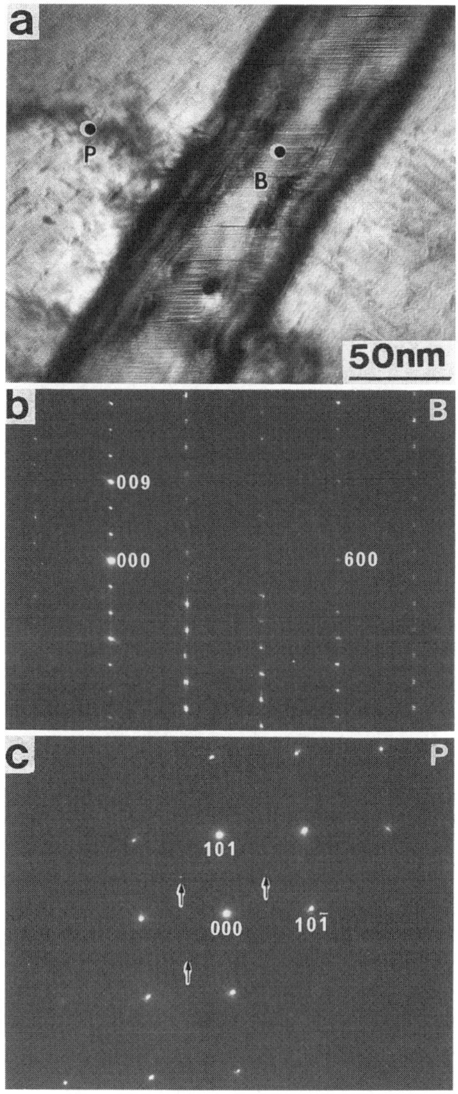

Figure 11. TEM image of a bainite plate of about 50 nm thick in an early stage of growth, observed in a specimen heat-treated at 433 K for 5.4 ks, (a), and ED patterns taken from the areas indicated by black spots in the plate and the adjacent matrix in (a), (b) and (c), respectively.

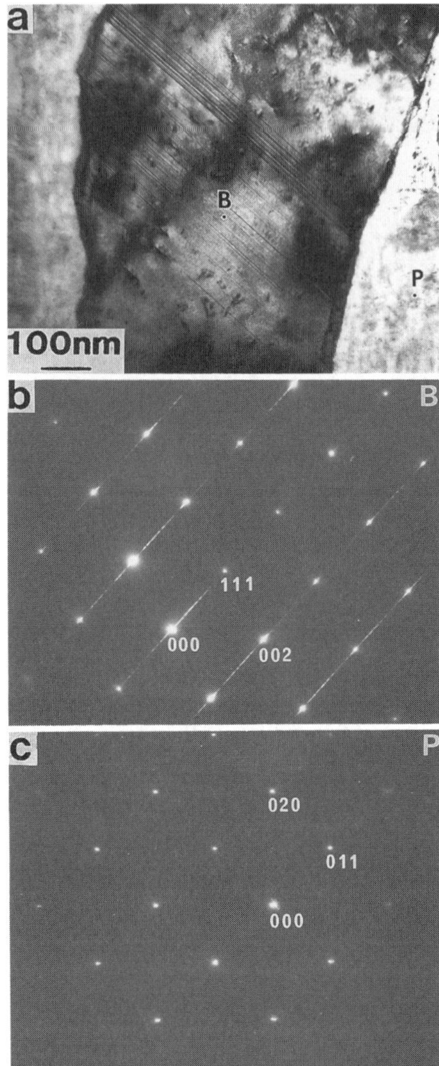

Figure 12. TEM image of a bainite plate of several hundreds nm thick in a later stage of growth, observed in a specimen heat-trated at 433 K for 1468.8 ks, (a), and ED patterns taken from the areas indicated by black spots in the plate and its adjacent matrix in (a), (b) and (c), respectively.

Figure 13 is a series of TEM image and EDX spectra, which were taken from a bainite plate of 30 nm thick and its adjacent matrix in a specimen heat-treated at 433 K for 2.4 ks. The black spots B and P in the image indicate the areas from which EDX spectra (b) and (c) have been taken, corresponding to the bainite and its adjacent matrix, respectively. This series thus give us a compositional information on an early stage of bainite growth. Another similar series for a bainite plate of 150 nm thick has been taken from the same specimen in order to obtain a compositional information on a middle stage of bainite growth. Both the series showed that Cd content in the bainites of early and middle stages of growth are similarly lower than that of the adjacent matrices.

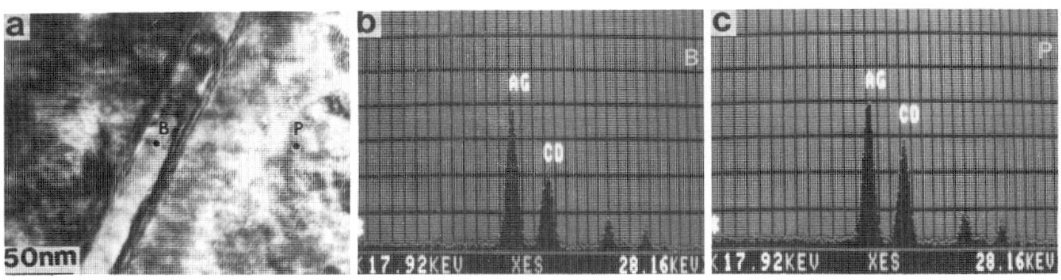

Figure 13. TEM image of a bainite plate in an early stage of growth, observed in a specimen heat-trated at 433 K for 2.4 ks, (a), and EDX spectra taken from the areas indicated by black spots in the plate and its adjacent matrix in (a), (b) and (c), respectively.

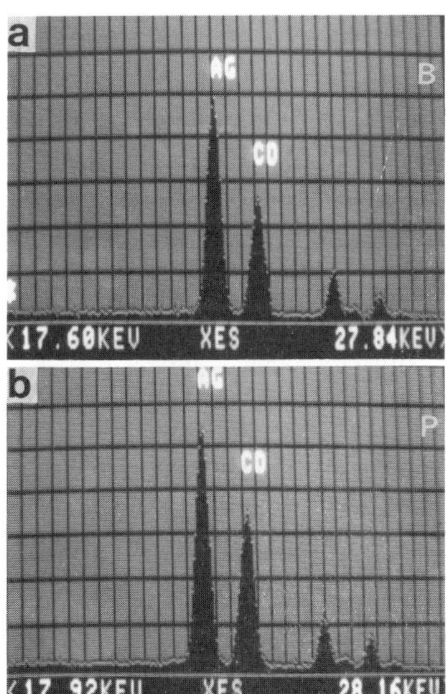

Figure 14. EDX spectra taken from the areas indicated by black spots in the bainite plate, (a), and its adjacent matrix, (b), in Figure 12(a).

Figure 14 shows EDX spectra, (a) and (b) being taken from a fcc bainite near the final stage of growth and its adjacent matrix, respectively. The spectra also show that Cd content is lower in the bainite plate than in the matrix. The Cd content thus analysed is shown in Figure 15, where the difference in Cd content between the bainite and matrix is plotted as a function of thickness of the bainite plates. The difference at the most right hand side is of the fcc α phase, and thus it is known from the figure that Cd content of the N9R bainite is nearly the same as that of the fcc α phase from the very early stage of growth. It may, therefore, be concluded that the 9R bainite plates are martensitically created in lower Cd content regions in the matrix, as in the case of Cu–Zn–A1 alloys, and they grow holding the reduced Cd content or changing a little the content at the growing interfaces due to a transformation stress, finally transforming to the fcc α phase.

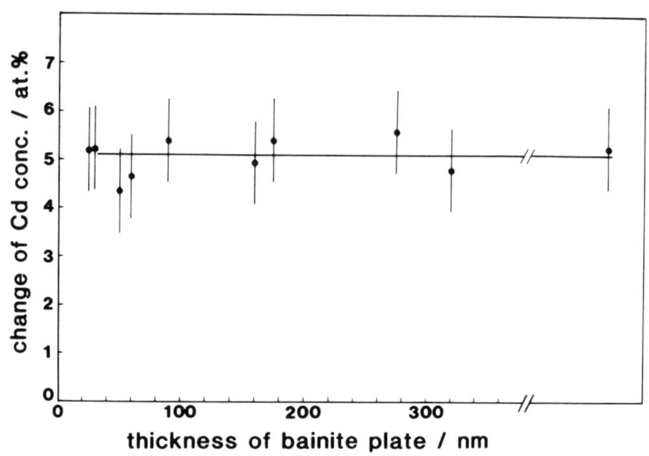

Figure 15. Deficiencies of solute Cd content in the bainite plates, compared with that in the surrounding matrix phase, as a function of plate thickness.

Internally faulted disordered N9R bainites have been observed in the Ag–Cd alloy, and they are obviously different from an internally twinned ordered 2H structure of the normal martensites produced in the subzero cooled alloy, being different from the case of Cu–Zn–A1 alloy. Such a modification from 2H to 9R martensites may come from a composition change in the matrix prior to the transformation, as has been observed in Cu–A1 alloys. That is, the Cu–A1 alloys with higher A1 content (26 to 28 at%) exhibit a martensitic transformation into 2H type γ_1', while those with lower A1 content (22 to 26 at%) do into 18 R type β_1'. Therefore, inversely speaking, the formation of 9R type martensite or bainite in the Ag–Cd alloy is a verification for the fact that the 9R type bainite is formed martensitically in lower Cd content regions.

3. ORDERING PROCESS AND MARTENSITIC TRANSFORMATION IN A Cu–Au–Zn ALLOY

The high temperature β phase of the alloy transforms into a $L2_1$ type ordered phase via a B2 ordered ordered phase at around 650 K upon cooling, and then the $L2_1$ phase further transforms into the martensite with a M18R type structure upon subsequent cooling. The shape memory effect and pseudoelasticity are known to be associated with the $L2_1$ to M18R martensitic transformation and its reversion. However, there has been an uncertainty as to site occupation of Cu and Zn atoms in the $L2_1$ parent phase and hence in the M18R martensite as well. Two structural models have so far been proposed as shown in Figure 16. The left hand side is by Duggin and Rachinger(9), and the right hand side is by Kubo and Shimizu(10). Our model was based on a hard sphere model for atoms, and was able to explain all crystallographic data without any contradiction, although it is inconsistent with compositional point of view.

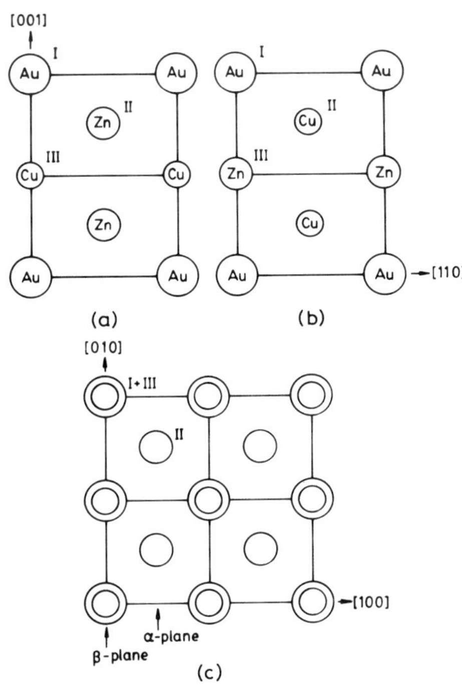

Figure 16 Atomic configurations on a {220} plane of the β phase Cu–Au–Zn alloy with the $L2_1$ structure proposed by Duggin and Rachinger, (a), and by Kubo and Shimizu, (b), and the (001) projection of the $L2_1$ structure, (c). The atomic sites I, II and III and the α- and β-planes are defined as shown.

The alloy used has the composition of Cu–20.7Au–43.8Zn (at%) as a result of chemical analysis. 3 mm diameter and 0.3 mm thick specimens with the $L2_1$ matrix, M18R martensites and aged martensites were supplied for ALCHEMI measurements and differential scanning calorimetry (DSC), and their powders were for X–ray diffractometry. The results obtained will be described for the $L2_1$ phase and as-quenched and aged martensites separately.

(1) $L2_1$ matrix phase [11]

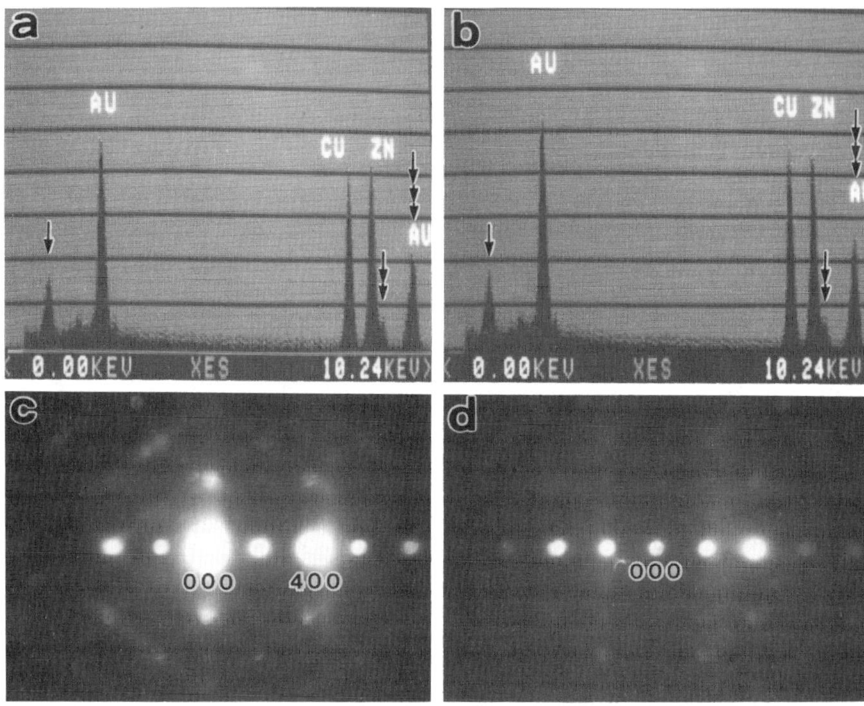

Figure 17. EDX spectra, (a) and (b), and ED patterns under which the spectra were acquired, (c) and (d). (a) and (c) are for $s > 0$ and (b) and (d) for $s < 0$ with respect to the 200 superlattice reflection. Single, double and triple arrows in (a) and (b) indicate the $Cu - L_\alpha$, $Cu - K_\beta$ and the mixture of the $Zn - K_\beta$ and $Au - L_\alpha$ spectra, respectively.

Atomic configuration in the $L2_1$ matrix phase at room temperature was examined by ALCHEMI under the planar channeling condition for a (200) plane. Figure 17 shows a series of EDX spectra and ED patterns taken from the $L2_1$ matrix of a Cu–Au–Zn specimen, (a) and (c) being the pair of them for a positive excitation error ($s > 0$) from the exact Bragg condition for the 200 superlattice reflection, and (b) and (d) being for a negative excitation error ($s < 0$). It is seen from the figure when the excitation error s changes from positive to negative, the intensity ratio of $Zn - K_\alpha$ to $Au - M_\alpha$ spectra appreciably decreases, while that of $Cu - K_\alpha$ to $Au - M_\alpha$ spectra remains almost constant. Such a change associated with diffraction condition in intensity of characteristic X–rays suggests that Au and Cu atoms lie on the same planes, while Au and Zn atoms do on different planes. Thus, we can know immediately that our model is not correct qualitatively.

Quantitative analysis of the ALCHEMI measurement was then carried out as follows: α– and β–planes are defined, as previously outlined, and Au atoms are assumed to not occupy the α-planes. Then k_{Cu} and k_{Zn} are introduced to express the fraction of Cu and Zn atoms occupying the β-planes, respectively. Accordingly, the distribution of all the constituent atoms into the α- and β-planes is expressed as shown in Table 2, where C_{Cu}, C_{Au} and C_{Zn} are atomic concentrations of Cu, Au and Zn atoms, respectively.

Table 2. Fractions of Cu, Au and Zn atoms on α- and β-planes, defined in Figure 16.

Atom	α-plane	β-plane
Cu	$C_{Cu} \cdot (1 - k_{Cu})$	$C_{Cu} \cdot k_{Cu}$
Au	0	C_{Au}
Zn	$C_{Zn} \cdot (1 - k_{zn})$	$C_{Zn} \cdot k_{Zn}$

Then, intensity of EDX spectrum emitted from an atom species i is given as follows:

(1) $$N_i^{(n)} = P_i \cdot C_i\{k_i \cdot I_\alpha^{(n)} + (1 - k_i) \cdot I_\beta^{(n)}\} ,$$

where P_i is a scalling factor related to X–ray fluorescent yield of i atoms, and $I_\alpha^{(n)}$ and $I_\beta^{(n)}$ are the electron beam intensities averaged over crystal thickness along the α- and β-planes, respectively. The superscript n represents diffraction condition, namely $n = 1$ is for positive s, $n = 2$ for negative s, and $n = 3$ for non-channeling condition. The occupation fractions k_{Cu} and k_{Zn} are not independent of, but are related to each other by the fraction of atom sites involved in α- and β-planes, as shown in the following equation:

(2) $$k_{Cu} \cdot C_{Cu} + C_{Au} + k_{Zn} \cdot C_{Zn} = 0.5.$$

From the above two equations, the occupation fraction of Cu atoms on β-planes is expressed by the following equation:

(3) $$k_{Cu} = \frac{R^{(1)} N_{Au}^{(1)}/N_{Au}^{(2)} - R^{(2)} \cdot N_{Cu}^{(1)}/N_{Cu}^{(2)}}{(1 - R^{(2)}) \cdot N_{Cu}^{(1)}/N_{Cu}^{(2)} - (1 - R^{(1)}) \cdot N_{Au}^{(1)}/N_{Au}^{(2)}} ,$$

where $R^{(n)}$ is the intensity ratio and shown as follows:

$$R^{(n)} = I_\alpha^{(n)}/I_\beta^{(n)}$$

$$(4) \qquad = \frac{N_{Zn}^{(n)} \cdot N_{Au}^{(3)}}{(1-k_{Zn}) \cdot N_{Zn}^{(3)} N_{Au}^{(n)}} - \frac{k_{Zn}}{1-k_{Zn}} .$$

Table 3. X–ray counts from the constituent atoms under the three kinds of diffraction conditions, and the results of analysis.

Sp. no	Atom	X–ray counts			$K_{Cu} \pm \Delta$
		$s > 0$	$s < 0$	Non-chan.	
1	Cu	9044	10614	11584	
	Zn	10368	10458	11120	0.94 ± 0.14
	Au	6300	7458	7744	
2	Cu	12404	12726	12884	
	Zn	13240	12084	11948	1.07 ± 0.24
	Au	8444	8610	8120	

Intensities of EDX spectra emitted from the constituent atoms were measured through the EDX spectroscopy, and k_{Cu} was determined from the intensities. Table 3 shows the X-ray intensities obtained from the EDX spectra shown in Figure 17 and another EDX spectra. k_{Cu} calculated from the intensities are also shown in the table to be almost unity. On the other hand, the maximum value of k_{Cu} is restricted by the following equation from the compositional point of view:

$$(k_{Cu})_{\max} = (\text{number of vacant sites on } \beta\text{-planes})/C_{Cu}$$
$$= 0.293/0.355 = 0.83.$$

Hence, the measured larger value of k_{Cu} means that the atom sites III on β-planes other than those (site I) occupied by Au atoms are completely taken over by Cu atoms. As a result, the distribution of all the constituent atoms into α- and β-planes is such as shown in Table 4. This result is also quantitatively consistent with the Duggin and Rachinger's model, but not with our model. As previously mentioned, our model was based on a hard sphere model for atoms, and explained well all the crystallographic data. Therefore, the above results suggest that atomic configuration in such a complicated structure is not necessarily determined by the simple hard sphere model.

Table 4. Site occupancies of Cu, Au and Zn atoms in the $L2_1$ phase of the Cu–Au–Zn alloy reduced from ALCHEMI measurements.

Atom	Occupancy fraction		
	α-plane	β-plane	
	II	I	III
Cu	0.062	0.043	0.25
Au	0.0	0.207	0.0
Zn	0.438	0.0	0.0

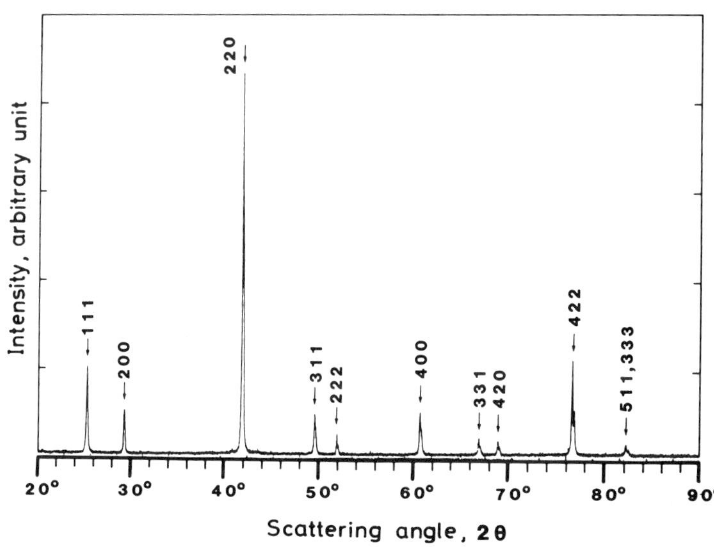

Figure 18. X–ray diffraction profile taken from a Cu–21Au–45Zn alloy powders at room temperature.

The ALCHEMI has revealed the fractions of site occupations of the constituent atoms on α- and β-planes. However, the distribution of Cu and Au atoms into site I and III on β-planes can not definitely be determined only by the ALCHEMI measurement. Then, powder X–ray diffraction has been carried out in order to examine the degree of order of next nearest neighbour atoms. Figure 18 shows a powder X-ray diffraction profile taken from the alloy at room temperature. All the reflection peaks are indexed by the $L2_1$ structure with the lattice constant of a=0.6070 nm.

Structure factor for the $L2_1$ lattice is written by the following equation:

$$
\begin{aligned}
F &= 4\cdot(f_I + 2f_{II} + f_{III}) = 4F_1 \quad \text{(for even hk1 and h+k+1 = 4n)}\\
&= 4\cdot(f_I - 2f_{II} + f_{III}) = 4F_2 \quad \text{(for even hk1 and h+k+1 = 4n+2)}\\
&= 4\cdot(f_I - f_{III}) = 4F_3 \quad \text{(for odd hk1)}\\
&= 0 \quad \text{(for mixed hk1)},
\end{aligned}
$$

where f_I, f_{II} and f_{III} are atomic scattering factors for the atoms located at sites I, II and III, respectively, and thus F_1 is the structure factor for the fundamental bcc lattice, and F_2 and F_3 are those for the nearest neighbour and next nearest neighbour orders in the $L2_1$ type superlattice, respectively.

Table 5. X–ray diffraction data and the results of analysis.

$2\theta°$	hkl	Type	Int.	$\lvert F_i\rvert_{obs.}$	$\lvert F_i\rvert_{cal.}$	$\lvert F_i\rvert_{obs.}/\lvert F_i\rvert_{cal.}$
42.0	220	F_1	100.0			
49.7	311	F_3	14.97	38.7	31.6	1.22
52.1	222	F_2	3.41	33.8	29.0	1.17
60.9	400	F_1	14.86			
67.1	331	F_3	5.34	33.5	29.0	1.16
69.1	420	F_2	4.31	31.2	26.2	1.19

Assuming that the present $Cu - Au - Zn$ alloy has a maximum degree of $L2_1$ order in the given composition, and that Zn atoms are all located at site II, as verified by ALCHEMI measurements, the structure factors F_2 and F_3 can be calculated from the site occupation previously determined by ALCHEMI, as shown in Table 5. Then, the ratios of the observed structure factors to the corresponding calculated ones becomes a measure for the degree of nearest neighbour and next nearest neighbour orders in the real alloy. The table also shows the results of analysis of the powder X-ray diffractometry. The ratios of $\lvert F_i\rvert_{obs}/\lvert F_i\rvert_{cal}$ are all close to unity. This indicates that not only the nearest neighbour ordering but also the next nearest neighbour ordering are completed in the present Cu–Au–Zn alloy during quenching. This result is in contrast with that in a $Cu - A1 - Ni$ alloy whose next nearest neighbour ordering was incompleted while the nearest neighbour ordering was completed when the alloy was rapidly quenched. Thus, in the $Cu - A1 - Ni$ alloy, M_s temperature was increased by about 50 K when slowly quenced, because of an increase in the degree of next nearest neighbour order. On the contrary, in the Cu–Au–Zn alloy, M_s temperature is hardly changed by quenching rate and subsequent aging at room temperature. This is consistent with the result obtained from the ALCHEMI and X-ray diffraction.

(2) As–quenched and ell-aged martensites[12]**.** In the Cu–Au–Zn alloy, the M18R martensite has been known to be stabilized by a martensite aging, bringing about an increase of A_s temperature, and the rubber-like behavior or twinning pseudoelasticity becomes to be realized in the alloy after the martensite aging. The stabilization and the rubber-like behavior are being paid attention from the standpoint of both practical application and fundamental understanding of shape memory alloys, but their origin has not been clarified yet. However, the origin is supposed to be related to a change of site occupation of the constituent atoms during the martensite aging. So, the atom configuration in the M18R martensite and well-aged one has been examined by the ALCHEMI and X-ray diffraction.

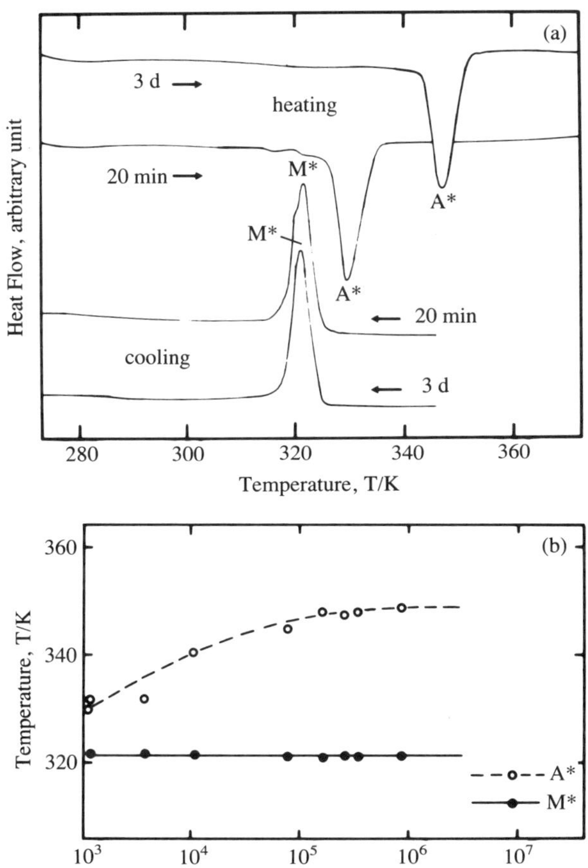

Figure 19. (a): DSC measurements of the $L2_1$ to M18R martensitic transformation temperatures in a Cu–26Au–45Zn (at%) alloy and their changes upon aging at room temperature, using an identical specimen. (b): Effects of aging at room temperature on the transformation temperatures, A^* and M^*, measured through the DSC.

Figure 19(a) shows differential scanning calorimetry (DSC) curves taken from an identical specimen aged at room temperature for 20 min. and 3 days after quenching. There can be seen that after aging for 3 days, A_s temperature rises by about 20 K while M_s temperature remains almost constant. The changing behavior of the A_s and M_s temperatures upon aging are shown in (b). A_s temperature rises steeply in the initial stage of aging and then levels off in a few days, while M_s temperature remains almost constant. Hence, the martensite of this alloy is known to be certainly stabilized during aging at room temperature, and the origin of the stabilization is removed completely once the alloy is heated above A_s temperature. (b) also indicates that the stabilization of martensite is saturated by aging for about 3 days, and a previous study(13) showed that an almost complete twinning pseudoelasticity was obtained for a specimen aged for more than 2 days at room temperature. So, ALCHEMI and X-ray diffraction were carried out for specimens aged for more than 2 days.

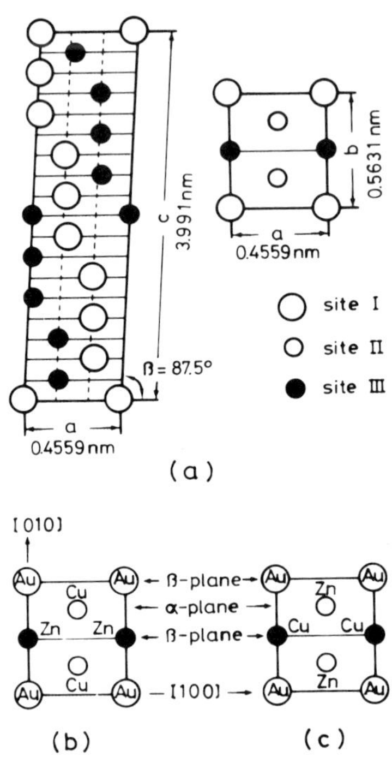

Figure 20. Crystal structure of the M18R martensite in a Cu–Au–Zn alloy near the stoichiometric composition $CuAuZn_2$, (a), and the basal planes proposed by Kubo and Shimizu, (b), and by Duggin and Rachinger, (c), the latter being derived from the $L2_1$ parent lattice following the lattice correspondence between the two lattices. The atomic sites I, II and III and the α- and β-planes are defined as shown in (b) and (c).

Figure 20 shows atomic configuration of the M18R martensite in the Cu–Au–Zn alloy and of the basal plane. (b) and (c) are of two models for the basal plane, as previously shown for the $L2_1$ matrix. α- and β-planes for the ALCHEMI measurements are also shown, which are parallel to the (020) plane. Figure 21 shows a TEM image of the inside of a M18R martensite and the corresponding ED patterns, surely showing the M18R type structure.

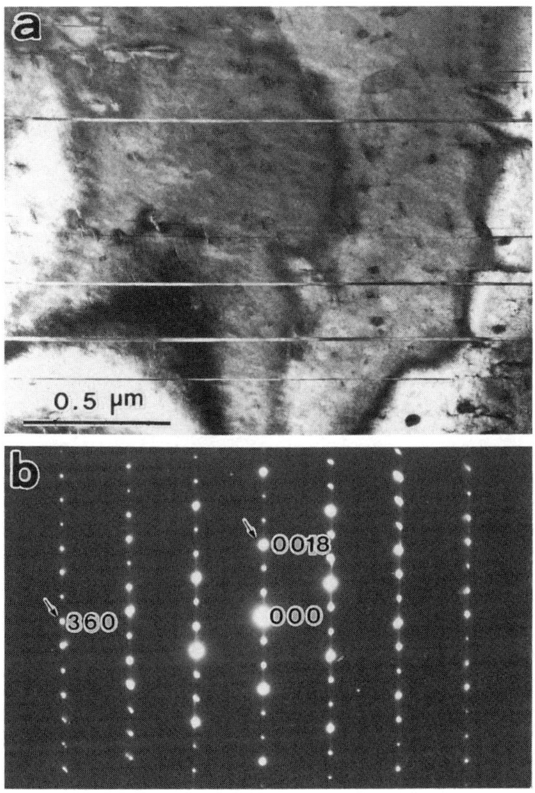

Figure 21. TEM image of the interior of a M18R martensite, (a), and the corresponding ED pattern, (b), showing that the M18R martensite of the present alloy is less faulted.

Figure 22 (a) and (b) are ED patterns showing that the area to be analysed by ALCHEMI is under the conditions of positive and negative excitation errors, respectively, with respect to the 020 superlattice reflection. (c) and (d) are EDX spectra obtained under the diffraction conditions corresponding to (a) and (b), respectively. According as the excitation error is positive or negative, intensity ratio of the $Zn - K_\alpha$ X–rays to the $Au - M_\alpha$ X–rays changes considerably, while that with respect to $Cu - K_\alpha$ and $Au - M_\alpha$ X–rays remains almost unchanged, similarly to the previous $L2_1$ matrix. The EDX spectra thus indicate clearly that Au and Zn atoms do not lie on the same planes, qualitatively.

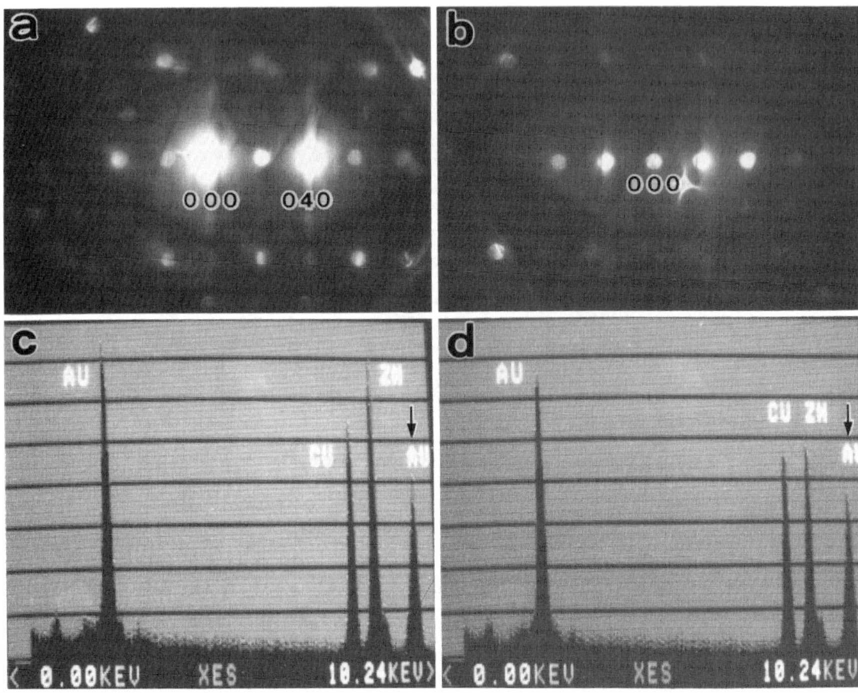

Figure 22. ED patterns, (a) and (b), under which EDX spectra, (c) and (d), were acquired. (a) and (c) are for $s > 0$, and (b) and (d) for $s < 0$ with respect to the 020 superlattice reflection. The arrows indicate spectra composed of $Au - L_\alpha$ and $Zn - K_\beta$.

Table 6. Fractions of Cu, Au and Zn atoms on α- and β-planes, defined in Figure 20.

Atom	α-plane	β-plane
Cu	$C_{Cu} \cdot (1 - k_{Cu})$	$C_{Cu} \cdot k_{Cu}$
Au	0	C_{Au}
Zn	$C_{Zn} \cdot (1 - k_{zn})$	$C_{Zn} \cdot k_{Zn}$

Then, quantitative analysis has been carried out for the above ALCHEMI measurement in the same way as for the $L2_1$ matrix. All Au atoms are again assumed to be located on β-planes but not on α-planes, and the occupation fraction of Cu and Zn atoms lying on β-planes are expressed as k_{Cu} and k_{Zn}, respectively, as in the case for the $L2_1$ matrix. Then, the distribution of the constituent atoms onto α- and β-planes is represented as shown in Table 6. C_{Cu}, C_{Au} and C_{Zn} are atomic concentrations of Cu, Au and Zn atoms, respectively. Intensities of characteristic X–rays from those atoms under three diffraction conditions are shown in Table 7 as to the EDX spectra in Figures 22(c) and (d) and another pair of EDX spectra. From the intensities, the fraction of Cu atoms on β-planes is calculated to be almost unity, as shown. Similarly to the previous analysis, the distribution of all the constituent atoms onto α- and β-planes is such as shown in Table 8. This result indicates that Zn atoms exist only on α-planes, but not on β-planes. No difference has been found in atomic configuration between unaged and well-aged martensites through the ALCHEMI measurement.

Table 7. X–ray counts from the constituent atoms under the three kinds of diffraction conditions, and the results of analysis.

Sp. no	Atom	X–ray counts			$K_{Cu} \pm \Delta$
		$s > 0$	$s < 0$	Non-chan.	
1	Cu	15400	20366	116696	
	Zn	18774	20698	18260	0.97 ± 0.09
	Au	12541	16664	13788	
2	Cu	12398	11142	6022	
	Zn	16552	11636	7388	0.94 ± 0.07
	Au	10222	9324	5630	

Table 8. Site occupancies of Cu, Au and Zn atoms in the M18R martensite of a Cu–26Au–45Zn alloy reduced from ALCHEMI.

Atom	Occupancy fraction		
	α-plane	β-plane	
	II	I	III
Cu	0.057	0.0	0.243
Au	0.0	0.25	0.007
Zn	0.443	0.0	0.0

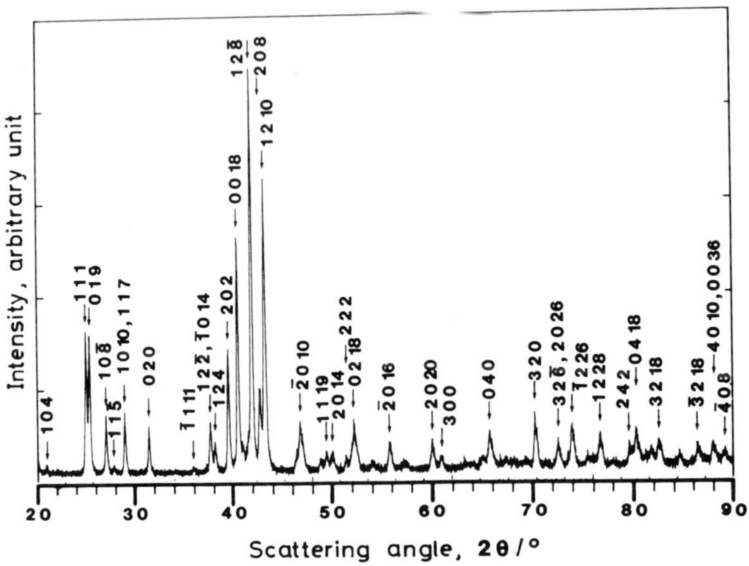

Figure 23. X–ray diffraction profile taken from the martensitic Cu–26Au–45Zn alloy powders at room temperature

On the other hand, the distribution of Cu and Au atoms into sites I and III could not be made clear by the ALCHEMI measurements as in the case of $L2_1$ matrix. So, this point has been examined by powder X–ray diffraction. Figure 23 shows an X–ray diffraction profile obtained from a martensitic specimen at room temperature. All the reflection peaks were able to be indexed by the M18R type structure with lattice parameters, $a = 0.4559$, $b = 0.5631$, $c = 3.991$ nm and $\beta = 87.5°$. Structure factor of the M18R martensite is written as follows:

$$F_{hk1} = F_A \cdot F_L,$$

where

$$
\begin{aligned}
F_A &= f_I + 2f_{II} + f_{III} = F_1 \quad \text{(for } k = 4n \text{ and even h, or k=4n+2 and odd h)},\\
&= f_I - 2f_{II} + f_{III} = F_2 \quad \text{(for } k = 4n \text{ and odd h, or k=4n+2 and even h)},\\
&= f_I - f_{III} = F_3 \quad \text{(for odd k)},
\end{aligned}
$$

and

$$F_L = \left[1+\exp\left\{2\pi i\left(\frac{h}{2.66}+\frac{k}{2}+\frac{\ell}{18}\right)\right\}+\exp\left\{2\pi i\left(\frac{2h}{2.66}+\frac{2\ell}{18}\right)\right\}\right]$$
$$\times\left[1+\exp\left\{2\pi i\left(\frac{h}{3}+\frac{k}{2}+\frac{3\ell}{18}\right)\right\}+\exp\left\{(2\pi i\left(\frac{2h}{3}+\frac{6\ell}{18}\right)\right\}\right]$$
$$\times\left[1+\exp\left\{2\pi i\left(\frac{k}{2}+\frac{9\ell}{18}\right)\right\}\right].$$

In these equations, F_A is the structure factor for the basal plane, and f_I, f_{II} and f_{III} are atomic scattering factors for the atoms located at sites I, II and III defined in Figure 20, respectively. Similarly to the previous analysis, the ratio $|F_2|_{obs}/|F_2|_{\mathrm{cal}}$ is a measure of the degree of order for ordering between the mixture of Cu and Au atoms at sites I and III, respectively, and Zn atoms at site II, and $|F_3|_{\mathrm{obs}}/|F_3|_{\mathrm{cal}}$ for ordering between Au atoms at site I and Cu atoms at site III. These ratios are shown in Table 9. According to the table, the degree of order with respect to the atomic distribution into the mixture of sites I and III and site II, that is, $|F_2|_{\mathrm{obs}}/|F_2|_{\mathrm{cal}}$, is almost unity, while that as to the atomic distribution into sites I and III, that is, $|F_3|_{\mathrm{obs}}/|F_3|_{\mathrm{cal}}$, is appreciably less than unity, and this is true for both the martensites aged for 2 days and 26 days. That is, the degree of order for ordering between Au atoms at site I and Cu atoms at site III is roughly 0.8, and then disordering occurs between Au and Cu atoms during aging at the martensitic state, since the ordering is completed in the $L2_1$ matrix phase as well as just after the martensitic transformation. It is then concluded that the origin of the stabilization of martensite and of the appearance of twinning pseudoelasticity is nothing but this type of disordering upon aging.

Table 9. X–ray diffraction data and the results of analysis.

$2\theta°$	hkl	Type	Int.		$\lvert F_i\rvert_{\mathrm{obs.}}$		$\lvert F_i\rvert_{\mathrm{cal.}}$	$\lvert F_i\rvert_{\mathrm{obs.}}/\lvert F_i\rvert_{\mathrm{cal.}}$	
			2d	26d	2d	26d		2d	26d
25.2	110	F_3							
			64.84	69.41	35.4	33.3	42.6	0.83	0.78
25.6	019	F_3							
31.5	020	F_2	9.40	9.73	41.6	39.7	41.4	1.00	0.96
42.0	128	F_1	100.0	100.0					

Acknowledgements. The author is grateful to Professor T. Tadaki and Messrs. Y. Nakata, T. Uyeda and H. Okazaki for their corporation in carrying out the above HRAEM and ALCHEMI experiments. The above studies were supported by a grant from the Research Program on "Creation of New Materials through Intelligent Design" of ISIR, Osaka University, and by the Grant-in-Aid for Fundamental Scientific

Research (Sogo A, 1989–90, Ippan B, 1990–91) of the Ministry of Education, Science and Culture, Japan. The support is greatly appreciated.

REFERENCES

[1] A. Koreeda, T. Ishibashi, K. Shimizu, M. Tomita, C. Kimura and H. Okano, Mater Charact., 25 (1990), 375.

[2] Special Issues (No. 4, 6 and 10) of Metall. Trans. A, (1990).

[3] I. Cornelis and C.M. Wayman, Scripta Metall., 7 (1973), 579.

[4] C.M. Lorimer, G. Cliff, H. Aaronson and K.R. Kinsman, Scripta Metall., 9 (1975), 271.

[5] Y. Nakata, T. Tadaki and K. Shimizu, Mater. Trans. JIM, 30 (1989), 107.

[6] G. Cliff and G.W. Lorimer, Proc. 5th, European Congress on Electron Microscopy, London (1972), p. 140.

[7] T. Tadaki, T. Uyeda and K. Shimizu, Mater. Trans. JIM, 30 (1989), 117.

[8] T. Tadaki, S. Hamada and K. Shimizu, Trans. Jpn. Inst. Metals, 18 (1977), 135.

[9] J.J. Duggin and W.A. Rachinger, Acta Metall., 12 (1964), 1015.

[10] H. Kubo and K. Shimizu, Trans. Jpn. Inst. Metals, 17 (1976), 330.

[11] T. Tadaki, H. Okazaki, Y. Nakata and K. Shimizu, Mater. Trans. JIM, 31 (1990), 935.

[12] T. Tadaki, H. Okazaki, Y. Nakata and K. Shimizu, Mater. Trans. JIM, 31 (1990), 941.

[13] S. Miura, S. Maeda and N. Nakanishi, Phil. Mag., 30 (1974), 565.

ON THE PROBLEM OF TWO WELLS

VLADIMÍR ŠVERÁK*

Abstract. We study the behaviour of energy-minimizing sequences of deformations of hyperelastic materials whose stored energy function has two invariant wells.

Key words. polyconvexity, quasiconvexity, rank-one convexity

1. Introduction. A mathematical model explaining microstructures observed in certain materials, which has been proposed in [2], is based on studying minimising sequences of variational integrals

$$I(\varphi) = \int_{\Omega} W(D\varphi(x))\, dx. \tag{1}$$

Here $\Omega \subset \mathbf{R}^n$ is the domain occupied by the reference configuration of the body, the mappings $\varphi : \Omega \to \mathbf{R}^n$ describe deformations of the body and W is the stored energy function of the given material. We assume that W is continuous and satisfies the frame indifference condition, i.e. $W(R \cdot X) = W(X)$ for each $X \in M^{n \times n}$ and each $R \in SO(n)$. We shall also assume that there is $p \geq n$ and $c > 0$ such that $W(X) \geq c|X|^p$ for large X. We can clearly change W by a constant without affecting the behaviour of the model. Therefore we can assume $W \geq 0$ and $K = \{X, W(X) = 0\} \neq \emptyset$. We shall be interested in the sequences φ_j of deformations whose energy $I(\varphi_j)$ approaches zero.

For many materials a reasonable assumption seems to be $K = SO(n)$, i.e. that $W(X) = 0$ if and only if X is an orientation-preserving isometry. In this case it is well-known that the deformations φ with $I(\varphi) = 0$ are isometries (and therefore are affine) and if φ_j converges weakly to φ in $W^{1,p}(\Omega)$ and $I(\varphi_j) \to 0$, then $D\varphi \equiv \text{const.} = X_0$ for some $X_0 \in K$ and $D\varphi_j \to X_0$ strongly in $L^p(\Omega)$. (See, for example, [5], [8].)

For some materials, however, the set K can be (at certain temperatures) more complicated and it becomes more difficult to describe the deformations with zero energy or the sequences of deformations whose energy approaches zero. In some cases these sequences oscillate and, as pointed out in [2], this may account for the observed microstructures.

In this note, which was motivated by [3], we want to look at the case $n = 2$ and K consisting of two orbits (called wells) of $SO(2)$, i.e. $K = SO(2) \cdot A \cup SO(2) \cdot B$ for some matrices A, B with $\det A > 0$ and $\det B > 0$. We also make some comments concerning the case $n = 3$.

For $n = 2$ we prove that if the two wells are not compatible, i.e. $\operatorname{rank}(X - Y) \geq 2$ for every $X, Y \in K$, $X \neq Y$, then the sequences of deformations whose energy

*Department of Mathematics, Heriot-Watt University, Edinburgh EH14 4AS (On leave from MFF UK, Sokolovská 83, Prague)

approaches zero exhibit the same behaviour as in the case of a single well. (See Theorem 2.)

The case of compatible wells has been studied in [3] for $n = 3$ under the additional assumption that $\det A = \det B$. In the case $n = 2$ we can remove this restriction and we give an explicit description of the set K^{qc} of all linear deformations Y for which there exist deformations with arbitrary small energy whose restriction to the boundary of Ω is Y. Under our assumptions the set K^{qc} coincides with the zero set of the quasiconvexification of W, which is the stored energy function of the relaxed functional related to the functional I. (See, for example, [4].) The deformations which can be approximated (weakly in $W^{1,p}$) by deformations of arbitrary small energy are characterized by $D\varphi \in K^{qc}$. The set K^{qc} is quite large. In the case $\det A = \det B = \delta$ it consists of the matrices with determinant δ which are contained in the convex hull of K. This might indicate the possibility that the material can almost flow like a liquid provided the gradients do not get out of the convex hull of K. If $\det A \neq \det B$ then K^{qc} typically has nonempty interior. Hence the class of deformations with small energy is very large indeed.

Each matrix $X \in K^{qc} \setminus K$ is the centre of mass of a (not necessarily unique) nontrivial Young measure which comes from a sequence of gradients and which is supported at K. In this situation we can say that K carries a (nontrivial) microstructure. If moreover X belongs to the zero set of the rank-one convexification of W, then these microstructures can be typically constructed as "layers within layers". (See [2], and [3].)

2. Preliminaries. In what follows we denote by Ω a bounded open connected subset of $\mathbf{R}^n$. The Lebesgue spaces $L^p(\Omega)$ and the Sobolev spaces $W^{1,p}(\Omega)$ are defined in the usual way.
We denote by $M^{m\times n}$ the set of all real $m \times n$ matrices. For $X \in M^{m\times n}$ we denote by $\hat{X}$ the minors of the matrix X. For $n = m = 2$ we have $\hat{X} = (X, \det X)$, for $n = m = 3$ we have $\hat{X} = (X, \operatorname{Cof} X, \det X)$, etc. In general $\hat{X} \in \mathbf{R}^N$ with $N = \sum_{k=1}^{\min(m,n)} \binom{m}{k}\binom{n}{k}$.
If K is a subset of $M^{m\times n}$, we let $\hat{K} = \{\hat{X}, \quad X \in K\}$.
For a subset S of a linear space we denote by $\operatorname{conv} S$ the convex hull of S.
Let $f : M^{m\times n} \to \mathbf{R}$ be a continuous function. The following conditions on f are related to weak lower semicontinuity of the integral $\int_\Omega f(Du(x))\,dx$. (See [1], [7].)
(i) f is *rank-one convex* if for each matrix $A \in M^{m\times n}$ and each rank-one matrix $B \in M^{m\times n}$ the function $t \to f(A + tB)$ is convex.
(ii) f is *quasiconvex* if for any matrix $A \in M^{m\times n}$ and any smooth function $\varphi : \Omega \to \mathbf{R}^n$ compactly supported in Ω the inequality $\int_\Omega f(A + D\varphi)\,dx \geq \int_\Omega f(A)\,dx$ holds true. The class of quasicovex functions is independent of Ω. (See [1], [7].)
(iii) f is *polyconvex* if $f(X) =$ convex function of minors of the matrix X, or, using the notation introduced above, if $f(X) = G(\hat{X})$ for some convex function $G : \mathbf{R}^N \to \mathbf{R}$.
It is well-known that (iii)$\Rightarrow$(ii)$\Rightarrow$(i). (See [1], [7].)

Let $K \subset M^{m\times n}$. We define various "semiconvex hulls" of K related to the classes of functions above. We set

$$\begin{aligned}
K^{rc} &= \{X \in M^{m\times n}, \quad f(X) \le \sup_{Y\in K} f(Y) \text{ for every rank-one convex } f : M^{m\times n} \to \mathbf{R}\} \\
K^{qc} &= \{X \in M^{m\times n}, \quad f(X) \le \sup_{Y\in K} f(Y) \text{ for every quasiconvex } f : M^{m\times n} \to \mathbf{R}\} \\
K^{pc} &= \{X \in M^{m\times n}, \quad f(X) \le \sup_{Y\in K} f(Y) \text{ for every polyconvex } f : M^{m\times n} \to \mathbf{R}\}.
\end{aligned}$$

Clearly $K^{rc} \subset K^{qc} \subset K^{pc}$. The set K^{pc} can also be described as the set of all matrices which cannot be separated from K by using minors. Of all the hulls introduced above K^{pc} is the easiest one to compute because of the following lemma.

LEMMA 1. *Let K be a compact subset of $M^{m\times n}$. Then*

$$K^{pc} = \{X, \quad \hat{X} \in \operatorname{conv} \hat{K}\}.$$

Proof. This follows easily from the definition of polyconvexity and from separation theorems for convex sets.

3. Two wells in two dimensions. Let $A, B \in M^{2\times 2}$ such that $\det A > 0$ and $\det B > 0$. We want to consider the problem mentioned in the introduction in the special case $n = 2$ and $K = SO(2)\cdot A \cup SO(2)\cdot B$. Considering transformations $\varphi \mapsto \tilde{\varphi}$ of functions, where $\tilde{\varphi}(\tilde{x}) = R\cdot\varphi(A^{-1}Q^{-1}\tilde{x})$, we see that we can assume without loss of generality that $A = I$ and $B = H = \left(\begin{smallmatrix}\lambda & 0\\ 0 & \mu\end{smallmatrix}\right)$, where $0 < \mu \le \lambda$. Exchanging A and B, if necessary, we can also assume that $\det H = \lambda\mu \ge 1$. We shall always assume this in what follows.

Let us first consider the case when the two wells are compatible, i.e. $\operatorname{rank}(X - Y) = 1$ for some $X, Y \in K$. Under our assumptions this is equivalent to $0 < \mu \le 1$, $\lambda > 1$. We introduce new coordinates $y = (y_1, y_2)$ and $z = (z_1, z_2)$ for matrices $X = \left(\begin{smallmatrix}x_{11} & x_{12}\\ x_{21} & x_{22}\end{smallmatrix}\right) \in M^{2\times 2}$ by

$$X = \begin{pmatrix} y_1 & -y_2\\ y_2 & y_1\end{pmatrix} + \begin{pmatrix} z_1 & -z_2\\ z_2 & z_1\end{pmatrix}\cdot\begin{pmatrix}\lambda & 0\\ 0 & \mu\end{pmatrix}. \tag{2}$$

We shall write $X = (y, z)$. We note that $R\cdot X = (Ry, Rz)$ for every $X \in M^{2\times 2}$ and every $R \in SO(2)$. We can write $K = \{(y, 0), |y| = 1\} \cup \{(0, z), |z| = 1\}$.

LEMMA 2. *Let $a \le b$ and let $\mathcal{K} \subset \mathbf{R}^n \times \mathbf{R}^n \times \mathbf{R}$ be given by $\mathcal{K} = \{(y, 0, a), |y| = 1\} \cup \{(0, z, b), |z| = 1\}$. Then*

$$\operatorname{conv}\mathcal{K} = \begin{cases} \{(y, z, a), \ |y| + |z| \le 1\} & \text{if } a = b\\ \{(y, z, t), \ |y| \le 1 - \frac{t-a}{b-a} \text{ and } |z| \le \frac{t-a}{b-a}\} & \text{if } a < b.\end{cases}$$

Proof. The statement is obvious for $n = 1$. The general case follows from the case $n = 1$ by considering transformations $(y, z, t) \mapsto (Ry, Qz, t)$ with $R, Q \in SO(n)$, which leave invariant all sets under consideration.

LEMMA 3. *Let $n = 2$ and let K be the two compatible wells above. Let (y,z) be the coordinates on $M^{2\times 2}$ given by (2). Then the polyconvex hull of K is given by*

$$
(3) \qquad K^{pc} = \begin{cases} \{X = (y,z),\ |y| + |z| \le 1 \text{ and } \det X = 1\} \text{ if } \det H = 1 \\ \{X = (y,z),\ |y| \le 1 - \frac{\det X - 1}{\det H - 1} \text{ and } |z| \le \frac{\det X - 1}{\det H - 1}\} \text{ if } \det H > 1. \end{cases}
$$

Proof. This follows easily from Lemma 1 and Lemma 2.

LEMMA 4. *Under the assumptions of Lemma 3 the rank-one convex hull of K coincides with the polyconvex hull. In other words, $K^{rc} = K^{pc}$.*

Proof. Since $K^{rc} \subset K^{pc}$ for any set K, it is enough to prove that $K^{pc} \subset K^{rc}$. The proof is based on a property of one-sheeted hyperboloids $\Sigma \subset \mathbf{R}^3$ which is well-known from elementary geometry: for any point of $a \in \Sigma$ there are exactly two lines contained in Σ passing through a. (For Σ_0 given by $x_1^2 + x_2^2 - x_3^2 = 1$ the lines contained in Σ_0 are given by $(\cos\alpha - t\sin\alpha, \sin\alpha + t\cos\alpha, t)$ and $(\cos\alpha - t\sin\alpha, \sin\alpha + t\cos\alpha, -t)$, $\quad t \in \mathbf{R}$. The general Σ can be transformed to Σ_0 by a suitable change of variables.)

Let us consider first the case $\det H = 1$. Let L be a linear subspace of $M^{2\times 2}$ defined by $L = \{X,\ x_{12} + x_{21} = 0\} = \{X = (y,z),\ z_2 = 0\}$. Since both K^{rc} and K^{pc} are clearly invariant under transformations $X \mapsto R \cdot X$ with $R \in SO(2)$ and each orbit of $SO(2)$ intersects L, it is enough to prove that $K^{pc} \cap L \subset K^{rc} \cap L$. Let R_1, R_2 be the two elements of $SO(2)$ for which $\det(H - R) = 0$ (for $\mu = 1$ we have $R_1 = R_2$) and let $\Sigma = \{X \in L, \det X = 1\}$. It is easily checked that Σ is a one-sheeted hyperboloid in L. We have $\operatorname{conv} K = \{X = (y,z),\ |y| + |z| \le 1\}$ and we can easily compute $\Sigma \cap \operatorname{conv} K$. This intersection consists of two connected surfaces (with boundary) S_1 and S_2, which are related by $S_2 = \{X,\ -X \in S_1\}$. We choose the notation so that $H \in S_1$. The boundary of the surface S_1 consists of the rank-one segments $[H, R_1]$ and $[H, R_2]$ and the arc $(R_1, R_2) \subset SO(2)$ containing I. Therefore every rank-one convex function which is ≤ 0 at K is also ≤ 0 at the boundary of the surface S_1. We have lines contained in Σ passing through each point of Σ. Any line contained in Σ must be a rank-one line and we see that any rank-one convex function which is ≤ 0 at the boundary of S_1 must be ≤ 0 at any point of S_1. Using a similar argument for S_2, we see that every rank-one convex function which is ≤ 0 at K is ≤ 0 at any point of $S_1 \cup S_2$. The proof of the case $\det H = 1$ is finished.

The proof of the general case is similar. Let f be a rank-one convex function which is ≤ 0 at K. To prove that $f \le 0$ at any point of K^{pc}, it is sufficient to prove that $f \le 0$ at the boundary of K^{pc}. For this it is enough to prove that $f \le 0$ at the sets $\tilde{F}_1 = \{X = (y,z),\ |y| = 1 - \frac{\det X - 1}{\det H - 1}\} \cap \operatorname{conv} K$ and $\tilde{F}_2 = \{X = (y,z),\ |z| = \frac{\det X - 1}{\det H - 1}\} \cap \operatorname{conv} K$. Using the transformations $X \mapsto R \cdot X$ with $R \in SO(2)$ in the same way as in the case $\det H = 1$ we see that it is enough to prove that $f \le 0$ at the sets $F_1 = \tilde{F}_1 \cap \{X = (y,z),\ y_2 = 0\}$ and $F_2 = \tilde{F}_2 \cap \{X = (y,z),\ z_2 = 0\}$. Each of these set consists of two connected surfaces with boundaries, which are contained in certain one-sheeted hyperboloids. (Now we have different hyperboloids for each of the two pieces.) Using the same argument as in the case $\det H = 1$ we see that $f \leq 0$ at $F_1 \cup F_2$. The proof is finished.

THEOREM 1. *Under the assumptions above we have $K^{rc} = K^{qc} = K^{pc}$. Moreover, K^{pc} is explicitly described by the formulae (3).*

Proof. This follows directly from the lemmas above.

Let us now consider the case of incompatible wells, i.e. the case when $\operatorname{rank}(X - Y) \geq 2$ for every $X, Y \in K, X \neq Y$. Under our assumptions this is equivalent to $1 < \mu \leq \lambda$. It is easily seen that in this case we have $K^{pc} = K$.

THEOREM 2. *Let $n = 2$ and let I be the functional (1). Suppose that W is continuous, nonnegative and that there is $p \geq n$ and $c > 0$ such that $W(X) \geq c|X|^p$ for large X. Suppose also that $\{X, W(X) = 0\} = K$ where K consists of the two incompatible wells above. Let $\varphi_j : \Omega \to \mathbf{R}^n$ be a sequence in $W^{1,p}(\Omega)$ such that $I(\varphi_j) \to 0$ and suppose that φ_j converges weakly to $\varphi \in W^{1,p}(\Omega)$. Then $D\varphi \equiv \text{const.} = X_0 \in K$ and $D\varphi_j \to X_0$ strongly in L^p.*

Proof. We observe that if the two wells are not compatible, then there is $\alpha > 0$ such that $\det(X - Y) \geq \alpha|X - Y|^2$ for every $X, Y \in K$. Let η be a nonnegative smooth function compactly supported in Ω. Our assumptions imply that we can write

$$
\begin{aligned}
0 = \int_\Omega & \det D(\eta(\varphi_j - \varphi_k))\,dx \geq \\
& \int_\Omega (-\eta|D\varphi_j - D\varphi_k||D\eta||\varphi_j - \varphi_k| + \eta^2 \det(D\varphi_j - D\varphi_k))\,dx \geq \\
& \int_\Omega (-\eta|D\varphi_j - D\varphi_k||D\eta||\varphi_j - \varphi_k| + \alpha\eta^2|D\varphi_j - D\varphi_k|^2)\,dx - \varepsilon_{jk} \geq \\
-\frac{1}{2\alpha} & \int_\Omega |D\eta|^2|\varphi_j - \varphi_k|^2\,dx + \frac{\alpha}{2}\int_\Omega \eta^2|D\varphi_j - D\varphi_k|^2\,dx - \varepsilon_{jk},
\end{aligned}
$$

where $\varepsilon_{jk} \to 0$ for $j, k \to \infty$. We see that the L^2-norm of $(D\varphi_j - D\varphi_k)$ on compact subsets of Ω is estimated by the L^2-norm of $(\varphi_j - \varphi_k)+$ a term approaching zero as $j, k \to \infty$. Since $\varphi_j \to \varphi$ strongly in L^2_{loc} by the imbedding theorems, we see that $D\varphi_j \to D\varphi$ strongly in L^2_{loc} and using our growth assumptions on W we see that in fact $D\varphi_j \to D\varphi$ strongly in $L^p(\Omega)$. Therefore $D\varphi \in K$ a.e. in Ω and it remains to prove that $D\varphi$ is constant. We first prove that $D\varphi \in W^{1,2}_{\text{loc}}(\Omega)$. Let $u \in \mathbf{R}^2$ and for small $h > 0$ let $\varphi_h(x) = (\varphi(x + hu) - \varphi(x))/h$. Let η be as above. For sufficiently small h we have

$$
\begin{aligned}
0 = \int_\Omega & \det D(\eta\varphi_h)\,dx \geq \int_\Omega (-\eta|D\varphi_h||D\eta||\varphi_h| + \eta^2 \det D\varphi_h)\,dx \geq \\
& \int_\Omega (-\eta|D\varphi_h||D\eta||\varphi_h| + \alpha\eta^2|D\varphi_h|^2)\,dx \geq \\
-\frac{1}{2\alpha} & \int_\Omega |D\eta|^2|\varphi_h|^2\,dx + \frac{\alpha}{2}\int_\Omega \eta^2|D\varphi_h|^2\,dx.
\end{aligned}
$$

We see that the L^2-norm of $D\varphi_h$ on compact subsets of Ω is estimated by the L^2-norm of φ_h . We can now use the well-known Nirenberg's Lemma to infer that $D\varphi \in W^{1,2}_{\mathrm{loc}}(\Omega)$. Since functions from $W^{1,2}$ are continuous on almost every line of any given direction and K consists of two connected components, we see that either $D\varphi \in SO(2)$ a.e. in Ω or $D\varphi \in SO(2)\cdot H$ a.e. in Ω. Changing variables, if necessary, we can assume that $D\varphi \in SO(2)$ a.e. in Ω. This implies that φ satisfies the Cauchy-Riemann conditions and therefore if can be viewed as a holomorphic function. The complex derivative φ' of this function satisfies $|\varphi'| = 1$ and therefore it must be constant. The proof is finished.

Remarks: 1. It is clear that Theorem 2 remains true for any number of mutually incompatible $SO(2)$-invariant wells, since the condition $\det(X - Y) \geq \alpha|X - Y|^2$ for every $X, Y \in K$ remains satisfied.
2. The case $n = 3$ is more difficult. We refer the reader to [3] for remarks concerning the case of compatible wells. I do not know whether Theorem 2 remains true for $n = 3$. One of the difficulties here is that for $n = 3$ the polyconvex hull K^{pc} of K may not coincide with K even though the two wells are incompatible. The characterization of those diagonal matrices H for which we have $K^{pc} = K$ does not seem obvious. I do not know whether or not in the case of incompatible wells and $n = 3$ we always have $K^{rc} = K$.

In certain cases it is possible to prove Theorem 2 for $n = 3$ by a similar method to that used above for $n = 2$: we can use the well-known fact that $\frac{\partial}{\partial x_j}(\operatorname{Cof} D\vartheta)_{ij} = 0$ in distributions for any function $\vartheta : \Omega \to \mathbf{R}^3$ belonging to the space $W^{1,2}(\Omega)$. Therefore we have

$$\int_\Omega (\operatorname{Cof} D\vartheta - \operatorname{Cof} D\tilde{\vartheta})_{ij} \frac{\partial}{\partial x_j}(\eta^2(\vartheta_i - \tilde{\vartheta}_i))\, dx = 0 \tag{4}$$

for every $\vartheta, \tilde{\vartheta}$ from $W^{1,2}(\Omega)$, where η is the usual cut-off function. If there is $\alpha > 0$ such that

$$(\operatorname{Cof} X - \operatorname{Cof} Y)_{ij}(X - Y)_{ij} \geq \alpha|X - Y|^2 \quad \text{for all } X, Y \in K, \tag{5}$$

we can proceed similarly as in the proof of Theorem 2 and prove the statement of this theorem for $n = 3$. It is not difficult to see that if $H = \operatorname{diag}(\lambda_1, \lambda_2, \lambda_3)$ where $\lambda_1, \lambda_2, \lambda_3$ are positive numbers, then (5) is satisfied for some $\alpha > 0$ if and only if

$$\sum_i \lambda_i + \sum_{j<k} \lambda_j \lambda_k < 3 + 3\lambda_1\lambda_2\lambda_3. \tag{6}$$

After this work was completed, I learned that similar results concerning incompatible wells were obtained by Matos in [6]. He also gives a more general condition than (6) under which $K^{pc} = K$.

Acknowledgement: I thank John Ball for several useful suggestions.

REFERENCES

[1] Ball, J.M., *Convexity conditions and existence theorems in nonlinear elasticity*, Arch. Rat. Mech. Anal., 63 (1978), pp. 337-403.

[2] Ball, J.M. and James, R.D., *Fine phase mixtures as minimizers of energy*, Arch. Rat. Mech. Anal., 100 (1987), pp. 13-52.

[3] Ball, J.M. and James, R.D., *Proposed experimental tests of a theory of fine microstructures and the two-well problem*, Preprint (1990).

[4] Dacorogna, B., *Direct Methods in the Calculus of Variations*, Springer, 1989.

[5] Kinderlehrer, D., *Remarks about equilibrium configurations of crystals.*, In J.M.Ball, editor, *Symp. Material Instabilities in Continuum Mechanics,* pp.217-242. Heriot-Watt, Oxford University Press, 1988.

[6] Matos, J.P., *Young measures and the absence of fine microstructures in the $\alpha - \beta$ quartz phase transition*, Preprint (1991).

[7] Morrey, Ch.B., *Multiple Integrals in the Calculus of Variations*, Springer, 1966.

[8] Reshetnyak, Yu.G., *Spaces of Mappings with Bounded Distortion*, AMS Provenience, 1989.

SOME REMARKS ON SEPARATELY CONVEX FUNCTIONS

LUC TARTAR*

I want to discuss here some results concerning separately convex functions. Most of these results were obtained some time ago but only mentioned to a few specialists, and I had not taken the time to publish them before, for obvious reasons. The motivation of these studies was nonlinear elasticity, but once I had solved an academic example where quasiconvexity had been replaced by separate convexity, it was not clear to me how to get further on. I find useful to choose this subject now in order to describe the evolution of some ideas during the last fifteen years.

YOUNG measures and separately convex functions

In July 1978, I gave a few lectures at Heriot-Watt University in Edinburgh, where I described some of my ideas about how to use the compensated compactness method, developed in collaboration with François MURAT, for attacking some difficult problems in Continuum Mechanics which interested me.

This set of lectures was part of a program organized by Robin KNOPS and John BALL and I am thankful to them for having given me the opportunity to describe some of my ideas which otherwise would not have appeared in print at that time. Knowing my habit of rarely writing down the notes corresponding to my lectures, John BALL had asked one of his students, Bernard DACOROGNA, to write these notes in my place [T].

As I lectured in parallel with Ron DiPERNA it also gave me the opportunity to discuss with him my ideas of a new attack on hyperbolic systems. I had already mentioned these ideas to Ron, but not described to him all the details of what I knew at the time, and it took a few years before he could apply successfully my method to hyperbolic systems of two equations.

The question of nonlinear elasticity was in my mind, even if not mentioned directly. However, my approach was quite different from that of John BALL [B], which extended the work of MORREY and was based on studying the sequentially weakly lower semi-continuous functionals, the so-called quasiconvex functions. Instead, I wanted to work with the equilibrium equations and try to identify which constitutive relations relating the stress to the strain were stable with respect to weak convergence. For that study, I advocated the old mathematical tool of Young measures together with the new tool of compensated compactness which gave constraints that these Young measures must satisfy. Not unrelated to this approach was my desire to develop a tool which could also be used for hyperbolic systems, because I wanted to understand the evolution of elastic bodies, and I was not entirely satisfied with the approach of minimizing energy functionals. Of course, my program was too ambitious as even now not so much has been done in that direction, but I had found an (academic) example where the Young measures could

*Carnegie-Mellon University

be characterized and, interestingly enough, this characterization contained more information than the knowledge of the list of all sequentially weakly lower semi-continuous functionals, which in my example were precisely all the functions of two independent variables which were separately convex. As we will see in a moment this counterexample was actually slightly misleading.

The basic way of using Young measures before the introduction of the compensated compactness method was to derive results like the following

PROPOSITION 1. *Let Ω be an open set of the plane and $\alpha, \beta, \gamma, \delta \in R$.*

*a) Let u_n and v_n be two sequences of $L^\infty(\Omega)$, satisfying $\alpha \leq u_n \leq \beta$ and $\gamma \leq v_n \leq \delta$ a.e. in Ω. Then there exists a subsequence indexed by m, and a measurable family $\nu_{x,y}$ of probability measures on $[\alpha, \beta] \times [\gamma, \delta]$ such that for every continuous function G on $[\alpha, \beta] \times [\gamma, \delta]$ the sequence $G(u_m, v_m)$ converges in $L^\infty(\Omega)$ weak * to the function g defined by $g(x, y) = \langle \nu_{x,y}, G \rangle$ a.e. $(x, y) \in \Omega$.*

*b) Let F be a convex function on $[\alpha, \beta] \times [\gamma, \delta]$. Let u_n, v_n be two sequences of $L^\infty(\Omega)$ satisfying $\alpha \leq u_n \leq \beta$ and $\gamma \leq v_n \leq \delta$ a.e. in Ω, converging to u_∞, v_∞ in $L^\infty(\Omega)$ weak *, with $F(u_n, v_n)$ converging to f_∞ in $L^\infty(\Omega)$ weak *. Then $f_\infty \geq F(u_\infty, v_\infty)$ a.e. in Ω.*

*c) Given a measurable family $\nu_{x,y}$ of probability measures on $[\alpha, \beta] \times [\gamma, \delta]$, there exists sequences u_n, v_n of $L^\infty(\Omega)$ satisfying $\alpha \leq u_n \leq \beta$ and $\gamma \leq v_n \leq \delta$ a.e. in Ω, such that for every continuous function G on $[\alpha, \beta] \times [\gamma, \delta]$ the sequence $G(u_n, v_n)$ converges in $L^\infty(\Omega)$ weak * to the function g defined by $g(x, y) = \langle \nu_{x,y}, G \rangle$ a.e. $(x, y) \in \Omega$.*

*d) Let F be a continuous function on $[\alpha, \beta] \times [\gamma, \delta]$. Assume that for every sequences u_n, v_n of $L^\infty(\Omega)$ satisfying $\alpha \leq u_n \leq \beta$ and $\gamma \leq v_n \leq \delta$ a.e. in Ω, converging to u_∞, v_∞ in $L^\infty(\Omega)$ weak *, with $F(u_n, v_n)$ converging to f_∞ in $L^\infty(\Omega)$ weak *, one can deduce $f_\infty \geq F(u_\infty, v_\infty)$ a.e. in Ω. Then F is convex.*

With the help of the compensated compactness method one could derive new results like the following

PROPOSITION 2. *Let Ω be an open set of the plane and $\alpha, \beta, \gamma, \delta \in R$.*

*a) Let u_n and v_n be two sequences of $L^\infty(\Omega)$, satisfying $\alpha \leq u_n \leq \beta$ and $\gamma \leq v_n \leq \delta$ a.e. in Ω and assume moreover that $\partial u_n / \partial x$ and $\partial v_n / \partial y$ are bounded in $L^\infty(\Omega)$. Then there exists a subsequence indexed by m, a measurable family $\mu^1_{x,y}$ of probability measures on $[\alpha, \beta]$, a measurable family $\mu^2_{x,y}$ of probability measures on $[\gamma, \delta]$, such that for every continuous function G on $[\alpha, \beta] \times [\gamma, \delta]$ the sequence $G(u_m, v_m)$ converges in $L^\infty(\Omega)$ weak * to the function g defined by $g(x, y) = \langle \mu^1_{x,y} \otimes \mu^2_{x,y}, G \rangle$ a.e. $(x, y) \in \Omega$.*

*b) Let F be a separately convex function on $[\alpha, \beta] \times [\gamma, \delta]$. Let u_n, v_n be two sequences of $L^\infty(\Omega)$ satisfying $\alpha \leq u_n \leq \beta$ and $\gamma \leq v_n \leq \delta$ a.e. in Ω, converging to u_∞, v_∞ in $L^\infty(\Omega)$ weak *, with $F(u_n, v_n)$ converging to f_∞ in $L^\infty(\Omega)$ weak *. Assume moreover that $\partial u_n / \partial x$ and $\partial v_n / \partial y$ are bounded in $L^\infty(\Omega)$. Then $f_\infty \geq F(u_\infty, v_\infty)$ a.e. in Ω.*

*c) Given a measurable family of probability measures μ_y^1 on $[\alpha,\beta]$ and a measurable family of probability measures μ_x^2 on $[\gamma,\delta]$, there exists sequences u_n, v_n of $L^\infty(\Omega)$ satisfying $\alpha \le u_n \le \beta$ and $\gamma \le v_n \le \delta$ a.e. in Ω, with $\partial u_n/\partial x$ and $\partial v_n/\partial y$ bounded in $L^\infty(\Omega)$ such that for every continuous function G on $[\alpha,\beta]\times[\gamma,\delta]$ the sequence $G(u_n,v_n)$ converges in $L^\infty(\Omega)$ weak * to the function g defined by $g(x,y) = \langle \mu_y^1 \otimes \mu_x^2, G\rangle$ a.e. $(x,y)\in\Omega$.*

*d) Let F be a continuous function on $[\alpha,\beta]\times[\gamma,\delta]$. Assume that for every sequence u_n, v_n of $L^\infty(\Omega)$ satisfying $\alpha \le u_n \le \beta$ and $\gamma \le v_n \le \delta$ a.e. in Ω, with $\partial u_n/\partial x$ and $\partial v_n/\partial y$ bounded in $L^\infty(\Omega)$, converging to u_∞, v_∞ in $L^\infty(\Omega)$ weak *, with $F(u_n,v_n)$ converging to f_∞ in $L^\infty(\Omega)$ weak *, one can deduce $f_\infty \ge F(u_\infty,v_\infty)$ a.e. in Ω. Then F is separately convex.*

Remark 3. Any condition like a) satisfied by all the Young measures which appear in a given problem gives a sufficient condition like b) describing some functions which are sequentially weakly lower semi-continuous under the conditions of the problem. Any information like c) about which Young measures can be constructed in a given problem gives a necessary condition like d) that all sequentially weakly lower semi-continuous functions under the conditions of the problem should satisfy.

The characterization of the set X of Young measures which appear in a given problem determines the list Y of all sequentially lower semi-continuous functions F for that problem: they are exactly the functions F satisfying the generalized Jensen's inequality $\langle \nu, F\rangle \ge F(\underline{\nu})$ for all $\nu \in X$, where $\underline{\nu}$ denotes the center of mass of ν.

On the contrary, the characterization of the set Y of all the sequentially lower semi-continuous functions F for a given problem does not characterize the set X of Young measures which appear in that problem. Certainly such a Young measure ν must belong to the set Z of probability measures satisfying the generalized Jensen's inequality $\langle \nu, F\rangle \ge F(\underline{\nu})$ for all $F \in Y$, but Proposition 2 gives an example where $X \ne Z$. Indeed the set of $\nu \in Z$ with $\underline{\nu} = a$ is convex while the set of $\nu = \mu^1 \otimes \mu^2 \in X$with $\underline{\nu} = a$ is not convex (if $\alpha \ne \beta$ and $\gamma \ne \delta$).

Remark 4. Under the hypotheses of Proposition 2, the measures $\mu_{x,y}^1$ and $\mu_{x,y}^2$ have some regular dependence with respect to (x,y). For example, for all functions F, ϕ and ψ of class C^1 with compact support, $\mu_{x,y}^1$ satisfies $|\int \langle \mu_{x,y}^1, F\rangle \phi'(x)\psi(y)dxdy| \le K max_{\alpha\le u\le\beta}|F'(u)| max_x|\phi(x)| max_y|\psi(y)|$, where K is a bound for $|\partial u_n/\partial x|$. This is the reason why μ^1 has been chosen independent of x and μ^2 independent of y in Proposition 2.c.

However, this counterexample is misleading for two reasons.

A first reason is that this distinction does not occur in the setting of nonlinear elasticity which had motivated the question. Indeed, as shown by David KINDERLEHRER and Pablo PEDREGAL [K&P], the Young measures associated to gradients are exactly the probability measures satisfying the generalized Jensen's inequality for all the quasiconvex functions.

A second reason is that the conditions $\partial u_n/\partial x$ and $\partial v_n/\partial y$ bounded in $L^\infty(\Omega)$

are not so natural in the context of the Calculus of Variations. John BALL and François MURAT have proposed instead to consider sequences of the form $u_n = \partial\phi_n/\partial y$, $v_n = \partial\psi_n/\partial x$ with ϕ_n and ψ_n converging in $W^{1,\infty}(\Omega)$ weak * and such that $\partial\phi_n/\partial x$ and $\partial\psi_n/\partial y$ are relatively compact in $L^2_{loc}(\Omega)$, or more generally sequences u_n, v_n converging in $L^\infty(\Omega)$ weak * with $\partial u_n/\partial x$ and $\partial v_n/\partial y$ belonging to a compact set of $H^{-1}_{loc}(\Omega)$. In that setting they have shown that Young measures are not necessarily tensor products [B&M]; we will come back to this question in a moment with a different approach.

On the consequences of separate convexity

In the Spring 1983, while we were visiting the Mathematical Sciences Research Institute in Berkeley, John BALL and Bob KOHN raised the following question concerning separately convex functions, which they were considering as a simple substitute for studying the difficult question of whether rank-one convex functions are quasiconvex, a question which has not been settled yet.

For a given subset A of the plane, find the largest subset B containing A such that all separately convex functions F satisfying $F \leq 0$ on A automatically satisfy $F \leq 0$ on B. Similarly, a finite set of points A_1, A_2, .., A_n being given in the plane, find all the points M such that there exist nonnegative coefficients θ_1, θ_2, .., θ_n with the property that $F(M) \leq \Sigma_j \theta_j F(A_j)$ for all separately convex functions F.

Let us assume first that the functions F are allowed to take the value $+\infty$, with the usual generalization of what convexity means. Assume moreover that the points A_1, A_2, .., A_n are such that no two of them are on the same horizontal ($y =$ constant) or vertical ($x =$ constant) line. In this case, no point M outside the given points A_j will satisfy such an inequality. Indeed, the function equal to 0 at each point A_j and $+\infty$ outside the given points is obviously separately convex and takes the value $+\infty$ at M.

Let us assume now that the functions F must be finite everywhere. If the points A_1, A_2, .., A_n are such that no two of them are on the same horizontal or vertical line, does there exist points M with nonnegative coefficients θ_1, θ_2, .., θ_n such that $F(M) \leq \Sigma_j \theta_j F(A_j)$ for all (everywhere finite) separately convex functions F?

I found that the answer depends upon the geometric configuration of the points A_j. For example

PROPOSITION 5. *In an euclidean plane, consider a square with vertices A_j and center C, the sides of the square having an arbitrary orientation with respect to the axes (assumed to be orthogonal). Then one has $F(C) \leq (\Sigma_j F(A_j))/4$ for all (everywhere finite) separately convex functions F.*

Remark 6. In the case of 4 points A_1, .., A_4, a more general position is to consider a rectangle with vertices B_1, .., B_4 and sides parallel to the axes and to construct the points A_j in the prolongation of the sides of the rectangle taken in cyclic order. For example, let $x_1 < x_2$, $y_1 < y_2$ and $B_1 = (x_1, y_1)$, $B_2 = (x_2, y_1)$,

$B_3 = (x_2, y_2)$ and $B_4 = (x_1, y_2)$. For z_1, z_2, z_3, $z_4 > 0$, one configuration is $A_1 = (x_2 + z_1, y_1)$, $A_2 = (x_2, y_2 + z_2)$, $A_3 = (x_1 - z_3, y_2)$ and $A_4 = (x_1, y_1 - z_4)$. A second configuration, mirror image of the preceding one, is $A'_1 = (x_1 - z_1, y_1)$, $A'_2 = (x_2, y_1 - z_2)$, $A'_3 = (x_2 + z_3, y_2)$ and $A'_4 = (x_1, y_2 + z_4)$.

Let us assume that we are in the first configuration. Using convexity for each of the segments B_jA_j one obtains inequalities $F(B_{j+1}) \leq \eta_j F(B_j) + (1 - \eta_j)F(A_j)$ for some $\eta_j \in (0,1)$ which can be easily computed, and where as usual B_5 means B_1; one finds $\eta_j = z_j/(z_j + \Delta x)$ for $j = 1$, 3 and $\eta_j = z_j/(z_j + \Delta y)$ for $j = 2$, 4, where $\Delta x = x_2 - x_1$ and $\Delta y = y_2 - y_1$. One can easily eliminate all but one of the values $F(B_k)$ from these inequalities and deduce that for each j there are nonnegative coefficients $\theta_{j,k}$ depending only upon the coefficients η_m such that $F(B_j) \leq \Sigma_\kappa \theta_{j,k} F(A_k)$; for example $(1 - \eta_1\eta_2\eta_3\eta_4)F(B_1) \leq \eta_2\eta_3\eta_4(1 - \eta_1)F(A_1) + \eta_3\eta_4(1 - \eta_2)F(A_2) + \eta_4(1 - \eta_3)F(A_3) + (1 - \eta_4)F(A_4)$. Then one can easily deduce inequalities of the form $F(M) \leq \Sigma_j \theta_j F(A_j)$ for all (everywhere finite) separately convex functions F for any point M belonging to one of the segments $B_{j+1}A_j$ or to the closed rectangle $B_1B_2B_3B_4$.

Remark 7. For the points Mdescribed above, the coefficients θ_j can be obtained directly by solving the system of 4 equations $G(M) = \Sigma_j \theta_j G(A_j)$ where G is a basis of separately affine functions, e.g. $G = 1$, x, y and xy. For such points M, the coefficients θ_j are nonnegative (and have sum 1 by taking $G = 1$), as this is true for the points B_j. However, there are points M' for which the preceding system gives nonnegative coefficients θ_j with nevertheless no similar inequality $F(M) \leq \Sigma_j \theta_j F(A_j)$ holding for all (everywhere finite) separately convex functions F. In order to show that only the points M described above have this property, one uses some explicit separately convex functions F which are 0 at the points A_j and are arbitrarily large at the point M: a typical example of such a function is $C(x - a)_+(y - b)_+$ which is 0 for $x \leq a$ and $y \leq b$ and large when C goes to ∞ for some portion of the plane, other similar functions being derived by translation and symmetry from this one.

Remark 8. For a general distribution of points A_1, .., A_n, the preceding set of test functions modeled on x_+y_+ is not rich enough for excluding some points M for which no list of θ_j exists such that $F(M) \leq \Sigma_j \theta_j F(A_j)$ for all (everywhere finite) separately convex functions F. For example if $A_1 = (1,0)$, $A_2 = (0,1)$ and $A_3 = (-1,-1)$ it cannot eliminate the points M on the segments $[0, A_1)$ and $[0, A_2)$ for which no such inequality holds. The following result will correct this defect.

PROPOSITION 9. *Let A_i, $i = 1$, .., m be given in the quadrant $x > 0$, $y > 0$, and A_i, $i = m+1$, .., n be given in the quadrant $x \leq 0$, $y \leq 0$. If for a point M there exist nonnegative numbers θ_i, $i = 1$, .., n such that $F(M) \leq \Sigma_i \theta_i F(A_i)$ for all (everywhere finite) separately convex functions F, then either all the θ_i are 0 for $i = 1$, .., m or all the θ_i are 0 for $i = m+1$, .., n.*

Indeed if for example M belongs to the negative quadrant and all the points A_j, $j = 1$, .., m are in the region $xy \geq \epsilon > 0$ then one uses the separately convex function F defined by

a) $F(x,y) = 0$ if $x > 0$, $y > 0$ and $xy \geq \epsilon$

b) $F(x,y) = \epsilon$ in the negative quadrant

c) $F(x,y) = \epsilon - xy$ at all other points of the plane

and this shows that $\theta_j = 0$ for $j = 1, .., m$.

Remark 10. One can deduce that if a finite collection of points A_i is such that no two of them are on the same parallel to one axis and no four of them are in one of the positions described in Remark 6, then all these points are independent in the sense that for no point M outside these A_i there exist nonnegative numbers θ_i, $i = 1, .., n$ such that $F(M) \leq \Sigma_i \theta_i F(A_i)$ for all (everywhere finite) separately convex functions F.

We define $R = \{(x,y) | a \leq x \leq b; c \leq y \leq d\}$ to be the smallest rectangle containing all the points A_i. We rename the points so that $A_1 = (b, y_1)$ with the smallest possible y_1 and $A_3 = (a, y_3)$ with the largest possible y_3; eventually by using a symmetry, we can assume that $y_1 < y_3$. We rename the other points so that $A_2 = (x_2, d)$ with the smallest possible x_2 and $A_4 = (x_4, c)$ with the largest possible x_4. If it happened that A_2 and A_3 were identical, or that A_1 and A_4 were identical, it would be a corner and by Proposition 9 it would be independent from the other points so that one would be led to the same problem with less points and conclude by induction on the number of points. Because the four points do not satisfy the requirements of Remark 6, we deduce that $x_2 < x_4$. We define then by induction for $n \geq 3$, $A_{2n-1} = (x_{2n-1}, y_{2n-1})$ with $x_{2n-1} > x_{2n-2}$ and the largest possible y_{2n-1} and $A_{2n} = (x_{2n}, y_{2n})$ with $y_{2n} < y_{2n-1}$ and the smallest possible x_{2n}. One has $y_5 \geq y_1$ and then $y_{2n-1} \leq y_{2n+1}$ and similarly $x_{2n} \leq x_{2n-2}$ for $n \geq 3$. Because neither A_2, A_3, A_{2n-2}, A_{2n-1}, nor A_2, A_3, A_{2n-1}, A_{2n} satisfy the requirements of Remark 6, we deduce that $x_{2n} > x_2$ and $y_{2n-1} < y_3$ for $n \geq 3$. Because we only have a finite number of points, x_{2n} finally takes a constant value $x*$ and y_{2n-1} finally takes a constant value $y*$ and there are no points with $x > x*$ and $y > y*$ and no points with $y < y*$ and $x < x*$ so Proposition 9 enables us to split the set of points into two groups and one concludes by induction on the number of points.

Remark 11. If all the points are taken on the diagonal, then they are indeed independent and the preceding result suggests that the trace on the diagonal of (everywhere finite) separately convex functions can be arbitrary. Indeed for any smooth function g, there exists a separately convex function F, symmetric in (x,y), such that $F(x,x) = g(x)$ for all x.

Assuming that $g(x) = 0$ for $x \leq 0$, one defines F by $F(x,y) = g(y) + a(y)(x-y)$ for $x \leq y$ and $F(x,y) = g(x) + a(x)(y-x)$ for $x \geq y$, where the function a must be chosen adequately. The conditions that a must satisfy are $a'' \leq 0$, $2a' \leq g''$ and $2a \leq g'$ and one choice is to take $a(x) = 0$ for $x \leq 0$ and $2a'(x) = Inf_{0 \leq z \leq x} g''(z)$ for $x \geq 0$. One solves by symmetry the case where $g(x) = 0$ for $x \geq 0$ and by addition this answers the question for the case where g is a continuous function such that for every r there exists a constant $C(r)$ such that $g(x) + C(r)x^2$ is convex on $[-r, +r]$.

I do not know what is the precise class of functions g which can be the trace on the diagonal of (everywhere finite) separately convex functions. Vladimir ŠVERÁK has noticed that g'' cannot contain a negative Dirac mass and he recently communicated to me a remark of David PREISS giving a construction of a separately convex function whose second derivative is not bounded below on some small intervals.

Remark 12. There is no possible duality theory for separately convex functions, at least in the sense that we understand it for convex functions, as they are not bounded below by simple functions. The analog of polyconvex functions in our setting would be functions F given by $F(x,y) = \Phi(x,y,xy)$ where Φ is convex in its three arguments and such functions are automatically bounded below by $-C(1+|x|+|y|+|xy|)$ for some $C \geq 0$.

Remark 13. The method that I have used is not restricted to the case of separately convex functions and it is useful to extend it to a more general situation, while giving a differential version of what we have done before. Let E be a finite dimensional vector space and Λ a cone with vertex at 0. We will say that a function F is Λ-*convex*, if F is finite everywhere and $F(a+t\lambda)$ is convex in t for all $a \in E$ and all $\lambda \in \Lambda$. Let $M_1, .., M_n$ be some points in E and $e_1, .., e_n$ be some vectors in Λ; we want to consider the function f defined by $f(t) = \Sigma_i \theta_i(t) F(M_i + \alpha_i(t) e_i)$ and check under what choice of the functions θ_i and α_i, the function f is convex in t, whatever the Λ-*convex* function F is. By asking that the coefficient of $F''(M_i(t)) : (e_i, e_i)$ be nonnegative one gets $\theta_i {\alpha'_i}^2 \geq 0$; by asking that the coefficient of $F'(M_i(t)).e_i$ be zero one gets $2\theta'_i\alpha'_i + \theta_i\alpha''_i = 0$; by asking that the coefficient of $F(M_i(t))$ be zero, one gets $\theta''_i = 0$. This leads to $\theta_i(t) = \theta_i(0) + \kappa_i t$, $\alpha_i(t) = a_i t/(\theta_i(0) + \kappa_i t)$ as long as $\theta_i(0) + \kappa_i t > 0$. Notice that if F is Λ-*affine*, i.e. $\pm F$ is Λ-*convex*, then f is affine in t.

The condition $f'(0) \geq 0$ implies that f is nondecreasing for $t \geq 0$ (as long as $\theta_i(t) > 0$). A simple way to ensure this condition is to have $M_i = M_{i-1} + e_i$ for all i (with $M_{n+1} = M_1$ as usual) so that, by Λ-*convexity* of F, we have $F(M_{i-1}) \geq F(M_i) - F'(M_i).e_i$ for each i; as $f'(0) = \Sigma_i \theta'_i(0) F(M_i) + \theta_i(0)\alpha'_i(0) F'(M_i).e_i = \Sigma_i \kappa_i F(M_i) + a_i F'(M_i).e_i$ we deduce $f'(0) \geq \Sigma_i \kappa_i F(M_i) + a_i F(M_i) - a_i F(M_{i-1}) = 0$ if we impose $\kappa_i = a_{i+1} - a_i$ and $a_i \geq 0$. Notice that if F is Λ-*affine*, this choice gives f constant.

Assume now that $A_i = M_i + C_i e_i$ with $C_i > 0$ (the position of the points A_i described in Remark 6 corresponds to $n = 4$, the M_i having been denoted B_i there). If we want that at $t = 1$ the points $M_i(1)$ coincide with the points A_i, we have to choose the parameters $\theta_i(0) \geq 0$, $a_i \geq 0$ such that $\theta_i(0) + a_{i+1} - a_i > 0$ and $a_i/(\theta_i(0) + a_{i+1} - a_i) = C_i$. In the simple case where $\theta_1(0) = 1$ and $\theta_i(0) = 0$ for $i > 1$, this gives the system $a_1 = C_1(1 + a_2 - a_1)$, $a_i = C_i(a_{i+1} - a_i)$ for $i > 1$; this gives $a_j = \beta_j a_1$ for $j > 1$, where $\beta_j = \Pi_{j \leq k \leq n} C_k/(C_k + 1)$, and then $a_1 = C_1/(1 + C_1 - C_1\beta_2)$; this produces an inequality $F(M_1) \leq \Sigma_i \theta_i(1) F(A_i)$ valid for all Λ-*convex* functions F. The general case with $\theta_i(0) \geq 0$ for all i is obtained by a linear combination of these simple cases.

At another meeting organized at the Institute for Mathematics and its Applications in April 1985, I had shown examples of the above situation for plane nonlinear elasticity and also discussed of the relation with homogenization, and in particular the formula for iterated layering in its differential form which enables to use the principles of relaxed control. With the new tool of H-measures, which I had first developed for homogenization, one can expect to solve the question of whether all mixtures can be constructed by iterated layering and this could help decide about the relation of rank-one convexity to quasiconvexity. However, even more recently, it was not yet clear to me what my constructions about separately convex functions had to do with microstructures, because of that misleading result about Young measures being tensor products. Until Dick JAMES and Bob KOHN, led by some ideas of Graeme MILTON, used a similar construction for microstructures [B&F&J&K], and John BALL and François MURAT [B&M] pointed out that the Young measures would not be tensor products if $\partial u_n/\partial x$ and $\partial v_n/\partial y$ were only asked to stay in a compact of $H^{-1}_{loc}(\Omega)$.

Construction of YOUNG measures

We consider now sequences u_n, v_n converging to u_∞, v_∞ in $L^\infty(\Omega)$ weak * and satisfying $\alpha \leq u_n \leq \beta$, $\gamma \leq v_n \leq \delta$ a.e. in Ω and we only ask now that $\partial u_n/\partial x$ and $\partial v_n/\partial y$ belong to a compact of $H^{-1}_{loc}(\Omega)$. By the div-curl lemma, $u_n v_n$ converges to $u_\infty v_\infty$ in $L^\infty(\Omega)$ weak *, but as we cannot now obtain information on $\partial f(u_n)/\partial x$ or $\partial g(v_n)/\partial y$ we cannot learn anything on the limit of $f(u_n)g(v_n)$ and deduce, as in the case where $\partial u_n/\partial x$ and $\partial v_n/\partial y$ were bounded in $L^\infty(\Omega)$, that it is the product of the limits of $f(u_n)$ and $g(v_n)$ which implied that the Young measures were tensor products.

If a continuous function F on $[\alpha,\beta] \times [\gamma,\delta]$ is such that, under the preceding hypotheses on the sequences u_n, v_n, $F(u_n,v_n)$ converges to f_∞ in $L^\infty(\Omega)$ weak * implies $f_\infty \geq F(u_\infty,v_\infty)$ a.e. in Ω, then F must be separately convex. Indeed this will be consequence of the fact that one can obtain as Young measures the special measures $\delta_a \otimes (\theta\delta_c + (1-\theta)\delta_d)$ or $(\theta\delta_a + (1-\theta)\delta_b) \otimes \delta_c$, for all a, $b \in [\alpha,\beta]$, c, $d \in [\gamma,\delta]$ and $\theta \in [0,1]$. For example, the first one is obtained by taking $u_n = a$ and $v_n(x,y) = h(nx)$ where h is periodic with period 1 and defined by $h(t) = c$ for $0 \leq t < \theta$ and $h(t) = d$ for $\theta \leq t < 1$.

We want now to generate Young measures which are not tensor products, but we also want to generate as large a class as possible in order to decide if all separately convex functions have the property of sequential weak lower semi-continuity under our constraints. Taking a point $(u^*,v^*) \in [\alpha,\beta] \times [\gamma,\delta]$ we want to generate by induction some Young measures ν which are finite combinations of Dirac masses and have their center of mass at (u^*,v^*). If such a measure ν has been generated and contains a Dirac mass at (U,V) with a coefficient $C \geq 0$, we want to generate the measure ν' given by $\nu' = (C-D)\delta_U \otimes \delta_V + D\delta_U \otimes (\theta\delta_c + (1-\theta)\delta_d)$ with $0 \leq D \leq C$, $\theta \in [0,1]$ and c, $d \in [\gamma,\delta]$ such that $\theta c + (1-\theta)d = V$; similarly we want to generate the measure ν'' given by $\nu'' = (C-D)\delta_U \otimes \delta_V + D(\theta\delta_a + (1-\theta)\delta_b) \otimes \delta_V$ with $0 \leq C \leq D$, $\theta \in [0,1]$ and a, $b \in [\alpha,\beta]$ such that $\theta a + (1-\theta)b = U$.

All the measures ν generated in this way from the Dirac mass at (u^*,v^*) will have their center of mass at (u^*,v^*) because one does not change the center of mass by replacing $\delta_U \otimes \delta_V$ by $\delta_U \otimes (\theta\delta_c + (1-\theta)\delta_d)$ with $\theta \in [0,1]$ and $\theta c + (1-\theta)d = V$. All the measures ν generated in this way from the Dirac mass at (u^*,v^*) will also be such that $\langle\nu, F\rangle \geq F(u^*, v^*)$ for all separately convex functions; indeed, the value of $\langle\nu, F\rangle$ does not decrease when we split a Dirac mass in one of the two ways described above, i.e. $\theta F(U,c)+(1-\theta)F(U,d) \geq F(U,V)$ for $\theta \in [0,1]$ and $\theta c+(1-\theta)d = V$, and similarly $\theta \in [0,1]$ and $\theta a+(1-\theta)b = U$ imply $\theta F(a,V)+(1-\theta)F(b,V) \geq F(U,V)$.

We start from a constant function equal on Ω to (u^*, v^*) and decompose Ω into a disjoint union of rectangles R^k with sides parallel to the axes and of length $\leq \epsilon$, plus a set of measure $\leq \eta$; later, of course, ϵ and η will have to converge to 0. Suppose that we want to split $\delta_{u^*} \otimes \delta_{v^*}$ into $(1-D)\delta_{u^*} \otimes \delta_{v^*} + D\delta_{u^*} \otimes (\theta\delta_c + (1-\theta)\delta_d)$ with $0 \leq D \leq 1$, $\theta \in [0,1]$ and c, $d \in [\gamma, \delta]$ such that $\theta c+(1-\theta)d = v^*$; we first cut each of the rectangles R^k along a parallel to one of the axes into a rectangle R_1^k and a rectangle R_2^k, with $area(R_1^k) = (1-D)area(R^k)$ and $area(R_2^k) = Darea(R^k)$. On R_1^k we keep the preceding value (u^*, v^*) but on R_2^k we only keep u^* but change v^* into $v^* + \partial w/\partial x$ where w is a piecewise affine function constructed as follows.

Assume that the rectangle R_2^k is $(x_-, x_- + X) \times (y_-, y_- + Y)$ with X, $Y > 0$ and p is a large enough integer so that $(v^* - c)\theta X/p < Y/2$. The function w is described for $x_- \leq x \leq x_- + X/p$ and extended to R_2^k in order to be periodic in x with period X/p. In the trapezoid defined by $x_- \leq x \leq x_- + \theta X/p$ and $y_- + (v^* - c)(x - x_-) \leq y \leq y_- + Y - (v^* - c)(x - x_-)$, we take $w(x,y) = (v^* - c)(x - x_-)$. In the trapezoid defined by $x_- + \theta X/p \leq x \leq x_- + X/p$ and $y_- + (v^* - d)(x - x_- - X/p) \leq y \leq y_- + Y - (v^* - d)(x - x_- - X/p)$, we take $w(x,y) = (v^* - d)(x - x_- - X/p)$. In the triangle defined by $y_- \leq y \leq min\{y_- + (v^* - c)(x - x_-); y_- + (v^* - d)(x - x_- - X/p)\}$ we take $w(x,y) = y - y_-$. In the triangle defined by $max\{y_- + Y - (v^* - c)(x - x_-); y_- + Y - (v^* - d)(x - x_- - X/p)\} \leq y \leq y_- + Y$, we take $w(x,y) = -y + y_- + Y$.

The function w is 0 on the boundary of R_2^k and our constructions on R_2^k do not interfere with what we will do on other rectangles. The derivative $\partial w/\partial x$ is piecewise constant, equal to $v^* - c$ or $v^* - d$ in trapezoids with areas proportional to θ or $1 - \theta$ and equal to 0 in some small triangles along the horizontal sides of R_2^k. Similarly the derivative $\partial w/\partial y$ is 0 in the trapezoids and equal to -1 or $+1$ in the triangles. The areas of all the triangles in R_2^k is equal to $(v^* - c)\theta XY/p$ and can therefore represent a very small part of the area of R_2^k when p goes to infinity.

Because v is $v^* + \partial w/\partial x$ on R_2^k and w is 0 on the boundary of R_2^k, we can assume that w is extended to be 0 outside R_2^k and deduce that $\partial v/\partial y = \partial S/\partial x$ where $S = \partial w/\partial y$ on R_2^k, independently of what we do on the other rectangles R^j with $j \neq k$. As $\partial w/\partial y$ is ± 1 on triangles of total area $(v^* - c)\theta XY/p$, we see that the norm in H^{-1} of $\partial v/\partial y$ can be made as small as we want by taking p large enough on each of the rectangles where the construction is performed.

It is not difficult now to pass to the limit in the integral of $\phi(u,v)$ for any continuous function ϕ; any such sequence such that ϵ and η converge to zero and p converges to ∞ will generate the desired Young measure $(1-D)\delta_{u^*} \otimes \delta_{v^*} + D\delta_{u^*} \otimes (\theta\delta_c + (1-\theta)\delta_d)$.

It is not difficult to extend the preceding construction to a sequence of functions (u_n, v_n) which converges to a Young measure π which is a finite combination of Dirac masses. These functions will be assumed to be constant on each rectangle of a disjoint family of rectangles R^k with sides parallel to the axes and of length $\leq \epsilon$, these rectangles leaving aside a set of measure $\leq \eta$. The preceding argument gives a way to construct a new similar sequence (with the measure of the left aside set being slightly larger than η) converging to a Young measure obtained from π by splitting a Dirac mass of π into two Dirac masses in one of the two ways described before. We has therefore proved the following

PROPOSITION 14. *Let Ω be an open set of the plane and α, β, γ, $\delta \in R$. Consider sequences u_n and v_n of $L^\infty(\Omega)$, satisfying $\alpha \leq u_n \leq \beta$ and $\gamma \leq v_n \leq \delta$ a.e. in Ω, generating a Young measure ν , and such that $\partial u_n/\partial x$ and $\partial v_n/\partial y$ converge strongly to 0 in $H^{-1}_{loc}(\Omega)$.*

Among the possible Young measures ν which can be constructed in that way are the (x, y)-independent probability measures π on $[\alpha, \beta] \times [\gamma, \delta]$ belonging to the smallest set Z of probability measures on $[\alpha, \beta] \times [\gamma, \delta]$ satisfying the following requirements:

a) The measures $\delta_a \otimes (\theta\delta_c + (1-\theta)\delta_d)$ and $(\theta\delta_a + (1-\theta)\delta_b) \otimes \delta_c$ belong to Z, for all a, $b \in [\alpha, \beta]$, c, $d \in [\gamma, \delta]$ and $\theta \in [0, 1]$.

b) If μ is a nonnegative measure and $\mu + \kappa\delta_{u^*} \otimes \delta_{v^*}$ belongs to Z with $\kappa > 0$, then $\mu + \kappa\pi$ belongs to Z for any probability measure π having center of mass (u^*, v^*) and belonging to Z.

Remark 15. The smallest set Z satisfying a) and b) will consist only of combinations of Dirac masses at some special finite set of points. There are obviously always at least two of these points on the same horizontal or vertical line and therefore one cannot obtain a combination of Dirac masses at the points A_j, $j = 1, 2, 3, 4$, of the configuration described in Remark 6.

However one can obtain some combinations of Dirac masses using the points A_j and the points B_j for $j = 1, 2, 3, 4$ (under the condition that the center of mass belongs to the rectangle with vertices B_j). One can then split a Dirac mass at B_{k+1} into a Dirac mass at B_k with weight η_k and a Dirac mass at A_k with weight $1 - \eta_k$ and therefore render the weights of the points B_j, $j = 1, 2, 3, 4$, as small as desired. By a limiting argument, one can then obtain a combination of Dirac masses at the points A_j, $j = 1, 2, 3, 4$, alone.

This is in essence the argument of Dick JAMES and Bob KOHN for generating microstructures with a set without admissible connections [B&F&J&K]; the B_j are necessary for connecting the A_k together, but only an infinitesimal amount of them is used for that purpose so that the Young measure does not see the B_j.

It is then of importance to use the condition that the set Z be closed. Therefore one can add in Proposition 14 the requirement

c) Z is closed for the weak * topology.

Remark 16. The smallest set Z satisfying a) and b) obviously contains all the tensor products $\mu_1 \otimes \mu_2$ where μ_1, μ_2 are probabilities on $[\alpha, \beta]$ or $[\gamma, \delta]$ which are finite combinations of Dirac masses. Indeed, starting from the Dirac mass $\delta_{u^*} \otimes \delta_{v^*}$ and using successive splitting in the horizontal way, we can create $\mu_1 \otimes \delta_{v^*}$, where μ_1

is any probability which is a finite combination of Dirac masses and has its center of mass at u^*. Once Z is also required to satisfy c) it contains all the tensor products $\mu_1 \otimes \mu_2$ where μ_1, μ_2 are arbitrary probabilities on $[\alpha, \beta]$ or $[\gamma, \delta]$.

The smallest set Z satisfying a) and b) also satisfies a kind of convexity property. Indeed if π_1 and π_2 belong to Z and have the same center of mass (u^*, v^*), then $\theta\pi_1+(1-\theta)\pi_2$ belongs to Z for $\theta \in (0,1)$. Indeed it suffices to apply to $\theta\delta_{u^*}\otimes\delta_{v^*}$ the procedure creating π_1 by successive splitting of Dirac masses and to $(1-\theta)\delta_{u^*}\otimes\delta_{v^*}$ the similar procedure creating π_2. This property will obviously still hold once Z is required to satisfy c).

The problem is now to decide if the set Z contains all the probability measures ν satisfying $\langle \nu, F\rangle \geq F(\underline{\nu})$ for all separately convex function F. If it is so, this will be the characterization of the Young measures under the conditions that $\partial u_n/\partial x$ and $\partial v_n/\partial y$ are compact in H^{-1}_{loc}, a consequence being that separately convex functions are indeed sequentially weakly lower semi-continuous under these assumptions.

Questions of approximation

As a first step for attacking the preceding question, I think that it is useful to transform it into a finite dimensional question by an approximation argument. However, I do not know yet how to handle the discrete problem obtained.

Approximating separately convex functions is easy on a rectangular grid parallel to the axes, as one can use the usual Q_1 approximation of functions. When f is a continuous function on a rectangle R with sides parallel to the axes and vertices M_j, $j = 1, .., 4$, there is a unique interpolation function Πf on R belonging to Q_1, i.e. such that $\Pi f(M_j) = f(M_j)$, $j = 1, .., 4$ and of the form $f(u_1, u_2) = a + bu_1 + cu_2 + du_1u_2$. In order to simplify, let us consider a uniform square grid with mesh size $h = \Delta x = \Delta y$. For any function f defined in the whole plane and for all integers i, n, we denote $f_i^n = f(ih, nh)$ and let $\Pi_h f$ be the continuous piecewise Q_1 function interpolating f at all the grid points of the form (ih, nh).

Proposition 17. If f is a separately convex function defined in the whole plane, then $\Pi_h f \geq f$.

If f is a separately convex function defined in the whole plane, then $2f_i^n \leq f_i^{n-1} + f_i^{n+1}$ and $2f_i^n \leq f_{i-1}^n + f_{i+1}^n$ for all integers i, n.

Conversely if f is defined in the whole plane and satisfy $2f_i^n \leq f_i^{n-1} + f_i^{n+1}$ and $2f_i^n \leq f_{i-1}^n + f_{i+1}^n$ for all integers i, n, then $\Pi_h f$ is separately convex.

Let us show the first part for the unit square: for a separately convex function f and $0 \leq u_1, u_2 \leq 1$ we have $f(u_1,u_2) \leq (1-u_2)f(u_1,0)+u_2 f(u_1,1)$ by convexity in u_2, and as $f(u_1,0) \leq (1-u_1)f(0,0) + u_1 f(1,0)$ and $f(u_1,1) \leq (1-u_1)f(0,1) + u_1 f(1,1)$, one deduces that f is less or equal than a function in Q_1 which coincides with f at the four vertices of the unit square, i.e. $\Pi_1 f$.

The second part is easily seen by applying the convexity of f on the lines $u_1 = ih$ and on the lines $u_2 = nh$.

For the third part, let us consider the line $u_1 = ih$ and all the inequalities $2f_i^n \leq f_i^{n-1} + f_i^{n+1}$ for all integers n. As the restriction of $\Pi_h f$ to the line $u_1 = ih$

is piecewise affine on a uniform mesh, these inequalities simply express the fact that $\Pi_h f(ih, u_2)$ is convex in u_2. For $ih < u_1 < (i+1)h$, i.e. $u_1 = (1-\theta)ih + \theta(i+1)h$ with $0 < \theta < 1$, one has $\Pi_h f(u_1, u_2) = (1-\theta)\Pi_h f(ih, u_2) + \theta\Pi_h f((i+1)h, u_2)$ because $\Pi_h f(., u_2)$ is affine on the segment joining (ih, u_2) and $((i+1)h, u_2)$; this means that for every u_1, $\Pi_h f(u_1, .)$ is a convex combination of convex functions and is therefore a convex function. A symmetric argument using all the inequalities $2f_i^n \leq f_{i-1}^n + f_{i+1}^n$ for all integers i, n implies that $\Pi_h f(., u_2)$ is a convex function for all u_2.

As one can easily estimate the error between f and $\Pi_h f$ by using the modulus of uniform continuity of f, let us review first the implications of separate convexity of a function f upon its regularity. In one variable, it is easy to show that if a convex function g satisfies $g(\pm 1) \leq 1$ and $g(0) \geq 0$ then one has $-|t| \leq g(t) \leq 1$ and $|g'(t)| \leq (1+|t|)/(1-|t|)$ for $t \in (0,1)$. By using this argument on the line $u_1 = 0$, the line $u_2 = 0$, the lines $u_1 = t$ and the lines $u_2 = t$ one finds that if a separately convex function G satisfies $G(M) \leq 1$ for the four vertices of the square $(-1,+1)\times(-1,+1)$ and $G(0,0) \geq 0$ then one has $|u_1| - |u_2| - |u_1 u_2| \leq G(u_1, u_2) \leq 1$, $|\partial G/\partial u_1)| \leq (1+|u_1|)(1+|u_2|)/(1-|u_1|)$ and $|\partial G/\partial u_2)| \leq (1+|u_1|)(1+|u_2|)/(1-|u_2|)$ on the square $(-1,+1)\times(-1,+1)$. As a consequence, using $\gamma G(u_1, u_2) = f(Nu_1, Nu_2) - f(0,0)$, we deduce

PROPOSITION 18. *Let f be a separately convex function defined on the square $K_N = (-N,+N)\times(-N,+N)$ with $N > 0$, and let $\gamma = Max_j(f(M_j) - f(0,0))$ where the points M_j, $j = 1, .., 4$ are the four vertices of K_N. Then $Max_j f(M_j) - 4\gamma \leq f(u_1, u_2) \leq Max_j f(M_j)$ on K_N and $|\partial f/\partial u_1)| \leq 6\gamma/N$, $|\partial f/\partial u_2)| \leq 6\gamma/N$ on $K_{N/2}$.*

The basic error estimate for interpolation comes from the observation that $\Pi_h f = f$ if $f \in Q_1$ and that if $0 \leq f(M_j) \leq 1$, where M_j, $j = 1, .., 4$ are the four vertices of a square, then $0 \leq \Pi_h f \leq 1$ in this square. One deduces easily

PROPOSITION 19. *Let R be any square $(ih,(i+1)h) \times (nh,(n+1)h)$; if f is a Lipschitz function on R, then $|\Pi_h f(u_1,u_2) - f(u_1,u_2)| \leq Ch$ where C is an upper bound for $|\partial f/\partial u_1|$ and $|\partial f/\partial u_2|$ on R.*

Let f be a separately convex function and for $N > 0$, denote $\gamma(N) = Max_j(f(M_j) - f(0,0))$ where the points M_j, $j = 1, .., 4$ are the four vertices of K_N. Then $|\Pi_h f(u_1,u_2) - f(u_1,u_2)| \leq 3\gamma(2N)h/N$ on K_{N-h}.

Let us consider a family of real numbers f_i^n satisfying $2f_i^n \leq f_i^{n-1} + f_i^{n+1}$ and $2f_i^n \leq f_{i-1}^n + f_{i+1}^n$ for all integers i, n, and denote by F_h the piecewise Q_1 interpolation of these values, which is a particular separately convex function F_h satisfying $F_h(ih, nh) = f_i^n$ for all integers i, n. If one splits the mesh size in two or more, one needs more values for interpolation; a particular set of values is obtained by taking the values of F_h at the new points. If g is any other separately convex function satisfying $g(ih, nh) = f_i^n$ for all integers i, n, then $\Pi_h g = F_h$ and therefore Proposition 19 gives $|g - F_h| \leq Ch$ on K_{N-h} where C can be easily computed from the values f_i^n; for example C can be taken as $6Max|f_i^n|/N$ where the maximum is taken over the integers i, n, satisfying $-2(N+h) \leq i,\ n \leq 2(N+h)$. The uncertainty about the possible values that one can choose at one of the new points and still satisfy all the convexity requirements is then easily controlled.

Let us consider now a probability measure ν with compact support satisfying $\langle \nu, G\rangle \geq G(\underline{\nu})$ for all separately convex functions G, where $\underline{\nu}$ denotes the center of mass of ν. We want to approach ν by a probability measure ν_h which is a combination of Dirac masses at the points (ih, nh) for integers i, n, and we also want that ν_h still satisfy the same inequalities $\langle \nu_h, G\rangle \geq G(\underline{\nu}_h)$ for all separately convex functions G.

For this we notice that there is a unique probability $\nu_h^* = \Sigma_{i,n}\alpha_{i,n}\delta_{(ih,nh)}$ with compact support which satisfies $\langle \nu_h^*, \phi\rangle = \langle \nu, \Pi_h\phi\rangle$ for all continuous functions ϕ: indeed, one has $\langle \nu_h^*, \phi\rangle = \Sigma_{i,n}\alpha_{i,n}\phi(ih,nh) = \Sigma_{i,n}\alpha_{i,n}\Pi_h\phi(ih,nh) = \langle \nu_h^*, \Pi_h\phi\rangle$ and therefore, if $\psi_{i,n}$ denotes the piecewise Q_1 function which satisfies $\psi_{i,n}(jh,mh) = \delta_{ij}\delta_{mn}$ for all integers i, j, m, n, we find that the only solution is given by $\alpha_{i,n} = \langle \nu, \psi_{i,n}\rangle$ for all integers i, n. As $\psi_{i,n} \geq 0$, this gives $\alpha_{i,n} \geq 0$ for all integers i, n, and as $\Pi_h\phi = \phi$ if ϕ is affine, we deduce that ν_h^* is indeed a probability and that it has the same center of mass than ν.

In order to show that we can take $\nu_h = \nu_h^*$, we want to show that $\langle \nu_h^*, G\rangle \geq \langle \nu, G\rangle$ for all separately convex functions G, i.e. that $\langle \nu, \Pi_h G - G\rangle \geq 0$ for all separately convex functions G, and this is true as $\Pi_h G - G \geq 0$.

As h tends to 0 the sequence ν_h^* converges weakly * to ν, i.e. $\langle \nu_h^*, \phi\rangle$ converges to $\langle \nu, \phi\rangle$ for all continuous functions ϕ. Indeed, $\langle \nu_h^*, \phi\rangle - \langle \nu, \phi\rangle = \langle \nu, \Pi_h\phi - \phi\rangle$ and Proposition 19 tells us that $\Pi_h\phi - \phi$ converges uniformly to 0 on the support of ν if ϕ is a locally Lipschitz function and therefore that the same holds by density for all continuous functions ϕ.

As one can choose the origin at the center of mass of ν, it would be enough to characterize the probability measures ν of the form $\nu = \Sigma_{i,n}\alpha_{i,n}\delta_{(ih,nh)}$ with $\alpha_{i,n} \geq 0$, $\Sigma_{i,n}\alpha_{i,n}i = 0$, $\Sigma_{i,n}\alpha_{i,n}n = 0$, which satisfy $\Sigma_{i,n}\alpha_{i,n}f_i^n \geq f_0^0$ for all families f_i^n satisfying $2f_i^n \leq f_i^{n-1} + f_i^{n+1}$ and $2f_i^n \leq f_{i-1}^n + f_{i+1}^n$ for all integers i, n. The question is to decide if this set is exactly that of probability measures with center of mass at 0 which can be obtained as limits of sequences obtained from the Dirac mass at 0 by successive splittings as has been described earlier.

Acknowledgements

Most of my previously unpublished results described here had been obtained while I was visiting the Mathematical Sciences Research Institute in Berkeley or the Institute for Mathematics and its Applications in Minneapolis. My recent research work has been supported by the National Science Foundation under grant DMS-8803317.

REFERENCES

[B] John M. Ball, *Convexity conditions and existence theorems in nonlinear elasticity*, Arch. Rat. Mach. Anal., 63 (1977), pp. 337–403.

[B&M] John M. Ball & François Murat, *Remarks on rank-one convexity and quasiconvexity*, Proceedings, Dundee, July 1990 (to appear).

[B&F&J&K] Kaushik Bhattacharya & Nick Firoozye & Richard James & Robert V. Kohn, *Restrictions on microstructures*.

[K&P] David Kinderlehrer & Pablo Pedregal, *Characterizations of gradient Young measures*.

[T] Luc Tartar, *Compensated compactness and applications to partial differential equations*, in *Nonlinear Analysis and Mechanics: Heriot-Watt Symposium IV*, Pitman Research Notes in Mathematics 39, 1979, pp. 136–212.

NEW VARIATIONAL PROBLEMS IN THE STATICS OF LIQUID CRYSTALS

EPIFANIO G. VIRGA*

Introduction. Liquid crystals are anisotropic fluids that exhibit a preferred direction, possibly different at each place. The preferred direction is the *optical axis* of the material, that here we describe by a unit vector $\mathbf{n}$. Most liquid crystals present different phases as the temperature varies. We restrict attention to only one of these phases: our development concerns *nematic* liquid crystals. We assume the region of space occupied by a liquid crystal as given and we denote it by $\mathcal{B}$. Thus $\mathbf{n}$ is a field of $\mathcal{B}$ into $\mathbb{S}^2$, the unit sphere in the translation space of the three-dimensional Euclidean space $\mathcal{E}$. Formally,

$$\mathbf{n} : \mathcal{B} \to \mathbb{S}^2 \ , \mathcal{B} \subset \mathcal{E} \ , \ \mathbb{S}^2 := \{\mathbf{u} \in \mathcal{V} | \ |\mathbf{u}| = 1\}, \qquad \text{where } \mathcal{V} := \mathcal{E} - \mathcal{E}.$$

As $\mathcal{B}$ we take a *fit region*, that is a set which is regularly open, bounded, of finite perimeter, and of negligible boundary (see [1]). The optical axis is somehow prescribed on a part of the reduced boundary of $\mathcal{B}$:

$$\mathbf{n}|_D = \mathbf{n}_0, \quad \mathbf{n}_0 : D \to \mathbb{S}^2, \quad D \subset \partial^*\mathcal{B}. \tag{1}$$

The reduced boundary $\partial^*\mathcal{B}$ of the fit region $\mathcal{B}$ is the subset of the topological boundary $\partial\mathcal{B}$ where the outer unit normal $\boldsymbol{\nu}$ to $\mathcal{B}$ is defined. Various orientations of $\mathbf{n}$ can be induced on $\partial^*\mathcal{B}$ by *rubbing* it properly.

The stable equilibrium configurations of the optical axis in a liquid crystal minimize the *free energy*, that according to FRANK's theory [2] is the functional

$$\mathcal{F}_F[\mathbf{n}] := \int_{\mathcal{B}} \sigma_F(\mathbf{n}, \nabla\mathbf{n}), \tag{2}$$

subject to (1). In (2) σ_F is the real function defined by

$$\begin{aligned} \sigma_F(\mathbf{n}, \nabla\mathbf{n}) : = &\ k_1(\text{div } \mathbf{n})^2 + k_2(\mathbf{n} \cdot \text{curl } \mathbf{n})^2 + k_3|\mathbf{n} \wedge \text{curl } \mathbf{n}|^2 + \\ &+ (k_4 + k_2)(tr(\nabla\mathbf{n})^2 - (\text{div } \mathbf{n})^2), \end{aligned} \tag{3}$$

where k_1, k_2, k_3, and k_4 are constants.

If we set $k_1 = k_2 = k_3 = \kappa_F$ and $k_4 = 0$, σ_F reduces to

$$\sigma_F = \kappa_F|\nabla\mathbf{n}|^2. \tag{4}$$

*Facoltà di Ingegneria, Università di Pisa, 56126 PISA, ITALY

If $\partial\mathcal{B}$ is smooth and the boundary data for $\mathbf{n}$ are prescribed on the whole of $\partial\mathcal{B}$, i.e. if $D = \partial\mathcal{B}$ in (1), it was proved in [5] and [18] by HARDT, KINDERLEHRER and LIN that the functional $\mathcal{F}_F$ subject to (1) attains its minimum in a suitable class of admissible fields and, moreover, the minimizers are analytic away from a subset of $\mathcal{B}$ whose space dimension is less than 1. Thus the minimizers of $\mathcal{F}_F$ may possibly not be smooth mappings: in particular, they could be discontinuous somewhere in $\mathcal{B}$. The discontinuous minimizers of $\mathcal{F}_F$ are called *defects*. Defects which are discontinuous in isolated points are customary in liquid crystals. Also defects which are discontinuous in sets of space dimension equal or greater than 1 are observed, though they cannot be minimizers of FRANK's free energy. Such defects take different names according to the space dimension of the sets where their discontinuities occur. For example, there are *disclinations* and *textures* which are discontinuous on lines and surfaces, respectively. Disclinations are observed quite often in ordinary liquid crystals, while textures are just occasionally observed in polymeric liquid crystals (cf. e.g. [3]). A new mathematical theory to explain defects in a unified setting has recently been proposed by ERICKSEN in [4]. It is the aim of this paper to show by example the need for such a theory and the answers it provides to the problems that FRANK's theory left unsolved.

Disclinations. Let σ_F be given by (4). Suppose that $\mathcal{B}$ is a circular cylinder of heigh 2ℓ and radius R:

$$\mathcal{B} = \{p \in \mathcal{E} | -\ell < (p-o)\cdot\mathbf{e}_z < \ell, \quad 0 < (p-o)\cdot\mathbf{e}_r < R\}, \tag{5}$$

where o is a point of $\mathcal{E}$, $\mathbf{e}_z$ is a unit vector directed along the axis of the cylinder, and $\mathbf{e}_r$ is the radial unit vector orthogonal to $\mathbf{e}_z$. We denote by z and r the coordinates of $p \in \mathcal{B}$ along $\mathbf{e}_z$ and $\mathbf{e}_r$, respectively: $z(p) := (p-o)\cdot\mathbf{e}_z$, $r(p) := (p-o)\cdot\mathbf{e}_r$. We prescribe $\mathbf{n}$ on the lateral boundary of $\mathcal{B}$ as the unit outer normal:

$$\mathbf{n}|_{\partial_R\mathcal{B}} = \mathbf{e}_r, \quad \partial_R\mathcal{B} := \{p \in \partial\mathcal{B} | r = R\}, \tag{6}$$

and we leave $\mathbf{n}$ unprescribed on the bases of $\mathcal{B}$. We seek minimizers of $\mathcal{F}_F$ within the class of axisymmetric fields described by

$$\mathbf{n} = \cos\varphi(r)\mathbf{e}_r + \sin\varphi(r)\mathbf{e}_z, \tag{7}$$

where $\varphi : [0, R] \to \left[0, \frac{\pi}{2}\right]$ is a mapping subject to

$$\varphi(R) = 0.$$

One expects that the minimizer of $\mathcal{F}_F$ would be the field with $\varphi \equiv 0$, which indeed is discontinuous along the axis of the cylinder. Actually, FRANK himself proved that this field solves the Euler-Lagrange equations associated to $\mathcal{F}_F$, but it possesses *infinite* energy, despite to the fact that disclinations resembling it have been often observed. This puzzle was solved by CLADIS and KLÉMAN [6] who showed that the minimizer of $\mathcal{F}_F$ is a field that lies everywhere in the plane $(\mathbf{e}_r, \mathbf{e}_z)$, it is parallel to $\mathbf{e}_z$ only along the axis of the cylinder, and it is parallel to $\mathbf{e}_r$ only on the lateral boundary of the cylinder. Precisely, the angle φ is delivered by the function

$$\varphi(r) := \frac{\pi}{2} - 2\tan^{-1}\left(\frac{r}{R}\right), \quad r \in [0, R]. \tag{8}$$

Inserting (8) into (7) yields a *continuous* field. The disclination observed in experiments should be interpreted as a sign that $|\nabla \mathbf{n}|$ attains its maximum along the axis of the cylinder, though it is everywhere finite. All the experiments I refer to have been performed in capillary tubes (cf. e.g. [7], [8]).

Toward a new theory. The way toward a deeper understanding of disclinations in nematics did not end with the work of CLADIS and KLÉMAN. DE GENNES [9] proposed a theory to describe the change of liquid crystals into isotropic fluids that was modelled after the general theory of LANDAU for phase transitions. The *degree of orientation* s, a scalar that expresses how close a liquid crystal gets to its isotropic phase, came on the scene at that time. FAN [10] employed a degree of orientation variable in space to *tame* the singularities of $\mathbf{n}$: he allowed the liquid crystal to become isotropic wherever the classical theory predicts a singularity of $\mathbf{n}$ with infinite energy. ERICKSEN [4] has pursued FAN's line of thought and has set up a general theory for both the statics and dynamics of liquid crystals with variable degree of orientation.

The new theory. The scalar s ranges in the interval $\left[-\frac{1}{2}, 1\right]$. It measures the degree of orientation of the molecules that make up the material. The molecules are elongated in one preferred direction. The lower bound of s represents the state of microscopic order in which all molecules in a macroscopic site are orthogonal to $\mathbf{n}$, but are otherwise disordered. The upper bound of s corresponds to the state of perfect microscopic order in which all molecules are parallel to $\mathbf{n}$. Both these cases are ideal: they never occur in real specimens. In the isotropic phase of a liquid crystal $\mathbf{n}$ is deprived of any meaning: the molecules are completely disordered. The value $s = 0$ corresponds to such a microscopic state.

In the new theory s and $\mathbf{n}$ are fields defined thus:

$$s : \mathcal{B} \to [-1/2, 1], \quad \mathbf{n} : \mathcal{B} \setminus \mathcal{S}(s) \to \mathbf{S}^2,$$

where

$$\mathcal{S}(s) := \{p \in \mathcal{B} | s(p) = 0\}$$

is the *singular set* of the pair $(s, \mathbf{n})$. If $\mathcal{S}(s) \neq \emptyset$, we say that the pair $(s, \mathbf{n})$ is *singular*, otherwise we say that it is *regular*. The singular set of a pair $(s, \mathbf{n})$ is the set where discontinuities of $\mathbf{n}$ may occur. Thus, defects in liquid crystals are to be interpreted as *localized transitions to the isotropic phase.*

The free energy is the functional

$$\mathcal{F}_E[s, \mathbf{n}] := \int_{\mathcal{B}} \{\sigma_E(s, \nabla s, \mathbf{n}, \nabla \mathbf{n}) + \sigma_0(s)\}, \tag{9}$$

where σ_E and σ_0 are constitutive functions. I do not employ here the most general formula that ERICKSEN has proposed for σ_E in [4], I rather select a special instance of it that somehow parallels (4):

$$\sigma_E = \kappa_E\{k|\nabla s|^2 + s^2|\nabla \mathbf{n}|^2\}, \tag{10}$$

where κ_E and k are positive material moduli, both depending on the temperature T. Of course, when $s \equiv s_0$ (10) reduces to (4) with $\kappa_F = \kappa_E s_0^2$. The function σ_0 is often given the form

$$\sigma_0 = \frac{1}{2}as^2 - \frac{1}{3}bs^3 + \frac{1}{4}cs^4, \tag{11}$$

where a, b and c are non-negative moduli, of which only a depends on T. Equation (11) was derived by DOI [15] within his molecular theory for polymeric liquid crystals. The main features of σ_0 are illustrated in Figure 1 for five different values of T.

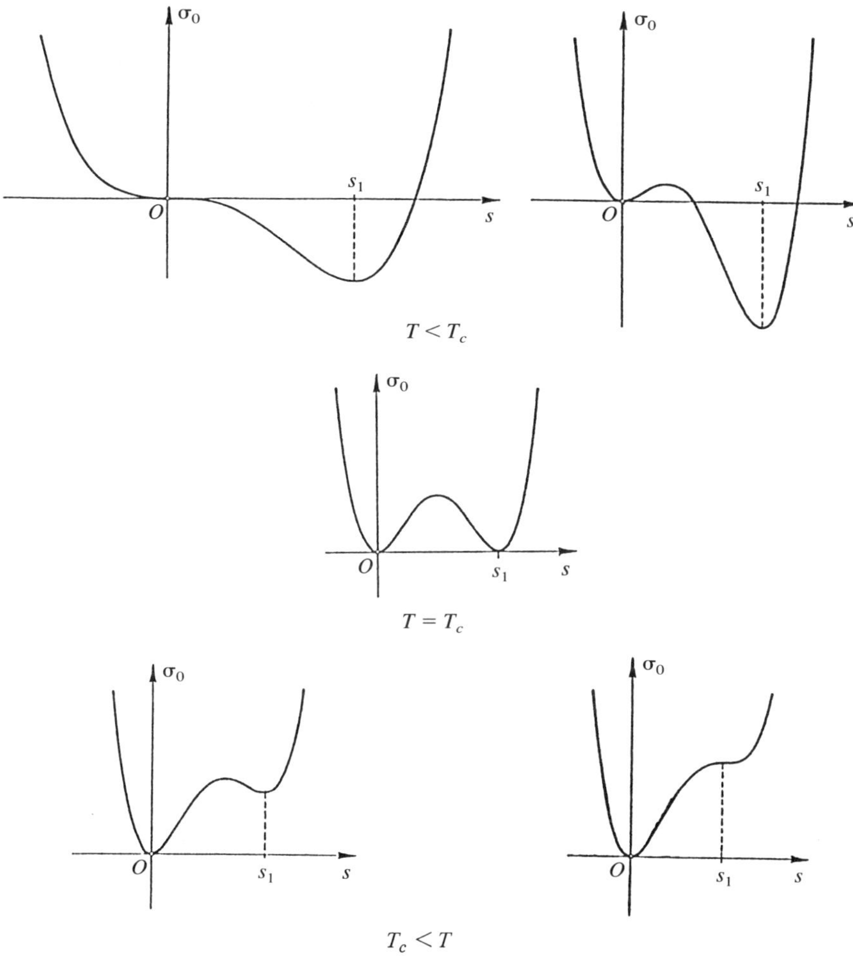

Figure 1

In the above graphs s_1 is a function of T: it is the degree of orientation of the equilibrium anisotropic phase, when both s and $\mathbf{n}$ are constant. The critical temperature T_c is the temperature above which s_1 is no longer the minimizer of σ_0.

Orders of magnitude. For ordinary liquid crystals there is experimental evidence that σ_0 prevails over σ_E by orders of magnitude (cf. [11]). Thus one should expect that the new theory systematically reduces to the classical one when $T < T_c$, since σ_0 is minimized only if $s \equiv s_1$. There are at least two objections to this conclusion. First, there is no evidence that σ_0 prevails over σ_E for polymeric liquid crystals, for which the new theory is also, if not principally, intended. Second, if T is sufficiently close to T_c, σ_0, though prevailing over σ_E, is unlikely to dictate the value of s everywhere, since $\sigma_0(0)$ and $\sigma_0(s_1)$ do not differ much.

In any event, we feel free to explore also the variational problems that arise when σ_0 is neglected in (9).

Boundary conditions. While several methods have been devised to induce a specific orientation of $\mathbf{n}$ on $\partial^*\mathcal{B}$, there is no way that I know of to prescribe s on $\partial^*\mathcal{B}$. The problems I will review here have been solved under different hypotheses, whose sole justification is their simplicity. Thus, disclinations have been accommodated into the new theory, by assuming that $s = s_1$ on $\partial^*\mathcal{B}$ or on part of it and by neglecting σ_0. Textures, which are not explained by the classical theory, are shown to fit into the new one, by employing a simplified formula for σ_0.

More about disclinations. Let $\mathcal{B}$ be the cylinder defined in (5). We prescribe $\mathbf{n}$ on the lateral boundary $\partial_R\mathcal{B}$ of $\mathcal{B}$ as in $(6)_1$ and moreover, we require that

$$s|_{\partial_R\mathcal{B}} = s_1, \tag{12}$$

where $\partial_R\mathcal{B}$ is defined as in $(6)_2$.

Let $\mathcal{F}_0$ be the functional defined by

$$\mathcal{F}_0[s,\mathbf{n}] := \int_{\mathcal{B}} \{k|\nabla s|^2 + s^2|\nabla\mathbf{n}|^2\}.$$

We call $\mathcal{R}$ the class of all pairs $(s,\mathbf{n})$ such that $s(p)$ depends only on $r(p)$ for all $p \in \mathcal{B}$ and $\mathbf{n}$ has the form (7).

THEOREM (MIZEL, ROCCATO and VIRGA [13]). *The pair of $\mathcal{R}$ that minimizes $\mathcal{F}_0$ subject to $(6)_1$ and (12) is singular for $k \leqq 1$ while it is regular for $k > 1$.*

Remark. The singular minimizer found in [13] is the pair

$$s(r) := s_1\left(\frac{r}{R}\right)^{1/\sqrt{k}}, \qquad \mathbf{n} = \mathbf{e}_r.$$

FRANK's original solution for disclinations is thus legitimate by the new theory. On the other hand, the regular minimizer has a field $\mathbf{n}$ that resembles the solution of CLADIS and KLÉMAN and a field s that decreases near the axis of the cylinder. Figure 2 illustrates the minimizing s for different values of k.

The graph in Figure 3 shows the minimum of $\mathcal{F}_0$ as a function of k. There the asymptothe corresponds to the value taken by $\mathcal{F}_0$ when $s \equiv s_1$ and $\mathbf{n}$ is the solution of CLADIS and KLÉMAN.

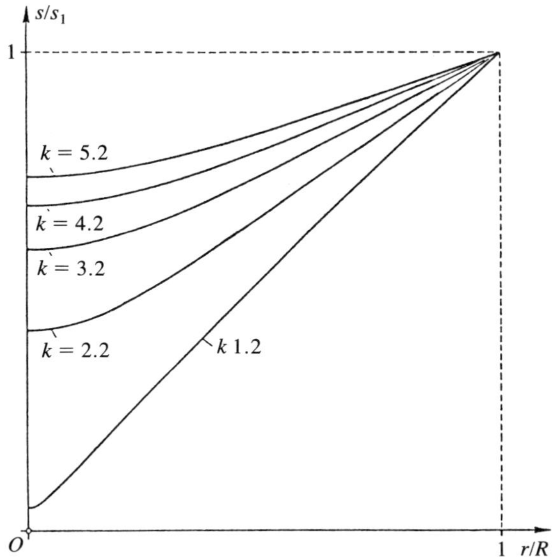

Figure 2

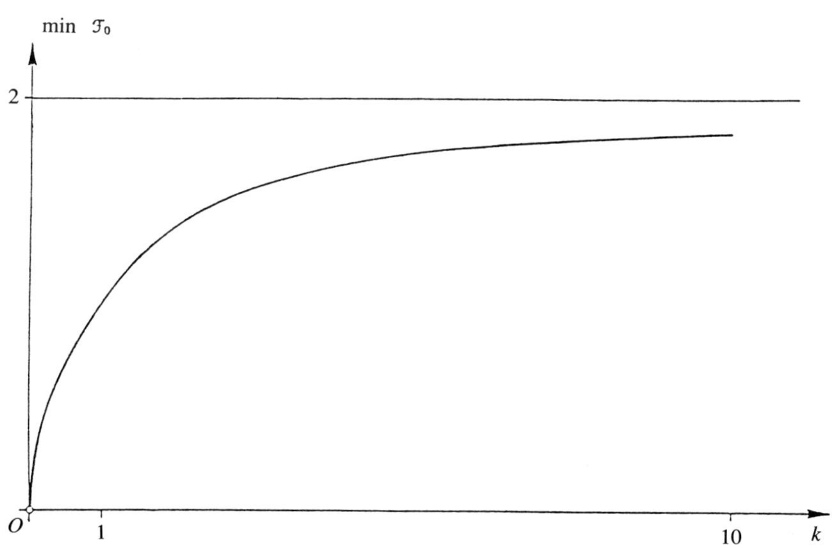

Figure 3

Textures. It has long been known that surface defects are not explained by the classical theory. In my opinion this has deferred researchers from interpreting the defects observed in some polymeric liquid crystal as textures. By a simple example we show here that textures can indeed minimize $\mathcal{F}_E$. Let $\mathcal{B}$ be the region between two parallel plates, 2ℓ apart:

$$\mathcal{B} = \{p \in \mathcal{E} | -\ell < (p-o)\cdot \mathbf{e} < \ell, \quad 0 < (p-o)\cdot \mathbf{e}_1 < \ell_1, \quad 0 < (p-o)\cdot \mathbf{e}_2 < \ell_2\},$$

where $o \in \mathcal{E}$ and $\mathbf{e}, \mathbf{e}_1, \mathbf{e}_2$ are mutually orthogonal unit vectors. We denote by x the coordinate of p along $\mathbf{e}$: $x(p) := (p-o)\cdot\mathbf{e}$. The plates that bound $\mathcal{B}$ are at $x = -\ell$ and $x = \ell$; we call them $\mathcal{S}^-$ and $\mathcal{S}^+$, respectively. There we prescribe $\mathbf{n}$ and s thus:

$$\text{(13)} \qquad \mathbf{n}|_{\mathcal{S}^-} = \cos\alpha_0\mathbf{e}_1 - \sin\alpha_0\mathbf{e}_2, \quad \mathbf{n}|_{\mathcal{S}^+} = \cos\alpha_0\mathbf{e}_1 + \sin\alpha_0\mathbf{e}_2, \quad s|_{\mathcal{S}^-\cup\mathcal{S}^+} = s_1,$$

where $\alpha_0 \in \left]-\frac{\pi}{2}, \frac{\pi}{2}\right[$ (see Figure 4).

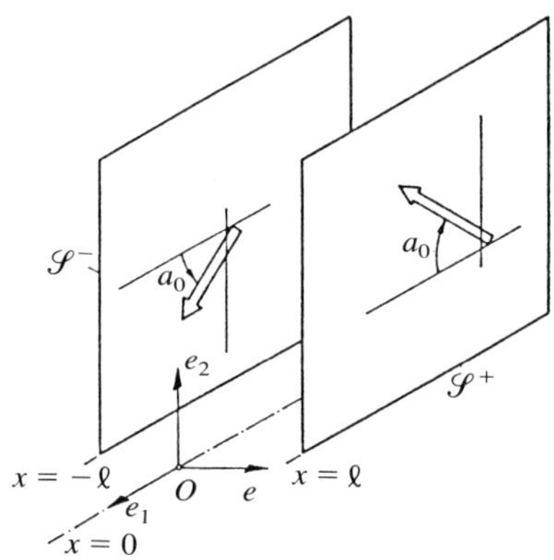

Figure 4

Suppose that $T < T_c$ in the potential σ_0 (cf. (11) and Figure 1).

THEOREM (VIRGA [12], ROCCATO and VIRGA [14]). *The functional $\mathcal{F}_E$ subject to (13) attains its minimum when*

$$\text{(14)} \qquad s(p) = \xi(x(p)) \ \textit{for all} \ p \in \mathcal{B}$$

and

$$\text{(15)} \qquad \mathbf{n}(p) = \cos\alpha(x(p))\mathbf{e}_1 + \sin\alpha(x(p))\mathbf{e}_2 \ \textit{for all} \ p \in \mathcal{B}\setminus\mathcal{S}(s).$$

It is also proved in [14] (cf. Section 3) that a minimizing s ranges in $[0, s_1]$, and so only the restriction of σ_0 to this interval really matters. We approximate σ_0 by a function that takes only three values: we set $\sigma_0(s) = \kappa_E \psi(s)$,

$$\psi(s) := \begin{cases} 0 & \text{for } s = 0, \\ h & \text{for } s \in \ \ ]0, s_1[, \, , \\ -a & \text{for } s = s_1, \end{cases}$$

where h and a are both positive constants. Our goal is to understand in a simple case how σ_0 interferes with occurrences of singular minimizers of $\mathcal{F}_E$.

We call $\mathcal{C}$ the class of all pairs $(s, \mathbf{n})$ that can be written as in (14) and (15) in terms of pairs (ξ, α) for which there is $\eta \in [0,1]$ such that

$$\xi(x) = s_1 \text{ for all } x \in [-\ell, -\eta\ell] \cup [\eta\ell, \ell].$$

(cf. Section 4 of [14] for more details on this class).

We set

$$z_c := \left(\frac{\pi\alpha_0 s_1}{4\ell}\right)^2$$

and we let k_c be the function defined by

$$k_c(z) := \begin{cases} \left(\frac{2\alpha_0}{\pi}\right)^2 & \text{for } z \in [0, z_c], \\ \left(\frac{\alpha_0^2 s_1}{2\ell}\right)^2 \frac{1}{z} & \text{for } z \in [z_c, \infty[. \end{cases}$$

THEOREM (ROCCATO and VIRGA [14]). *The minimizer in $\mathcal{C}$ of $\mathcal{F}_E$ subject to (13) is singular if, and only if,*

$$k \leqq k_c(h + a).$$

Remark. When $k \leqq k_c$ a surface defect develops at the mid plane between $\mathcal{S}^-$ and $\mathcal{S}^+$, the orientation becomes constant on both sides of the defect (cf. [14]). The texture consists of two adjacent *domains*.

Figure 5 shows the graph of k_c. We see that the parameters that really matter in deciding whether the minimizer of $\mathcal{F}_E$ in $\mathcal{C}$ in singular or not are k and $z := h + a$, the latter being proportional (for the special σ_0 employed here) to the depth of the potential well at $s = s_1$. It is worth noting that, no matter how big is z, the minimizer of $\mathcal{F}_E$ in $\mathcal{C}$ is singular for k sufficiently small. Furthermore, when $z \leqq z_c$, the critical value of k that somehow marks the border between regular and singular minimizers is the same as that found in [17], where we studied the functional $\mathcal{F}_0$, that is we took $\sigma_0 = 0$ in $\mathcal{F}_E$.

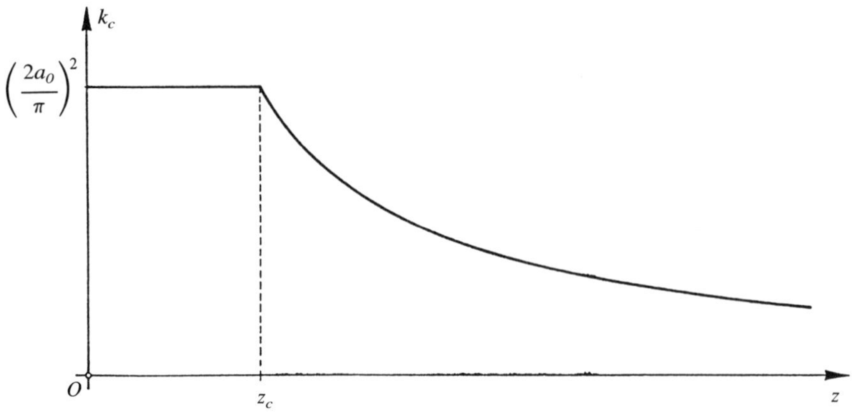

Figure 5

Acknowledgements. This paper was written while I was visiting the IMA at the University of Minnesota in October 1990. The support of the IMA is gratefully acknowledged. I thank Jerry Ericksen and Victor Mizel for interesting discussions on the matters reviewed in this paper.

REFERENCES

[1] W. Noll and E.G. Virga, *Fit regions and functions of bounded variation*, Arch. Rational Mech. Anal., **102** (1988), pp. 1–21.

[2] F.C. Frank, *On the theory of liquid crystals*, Discuss. Faraday Soc., **25** (1958), pp. 19–28.

[3] G.C. Berry, *Rheological properties of nematic solutions of rod-like polymers*, Mol. Cryst. Liq. Cryst., **165** (1988), pp. 333–360.

[4] J.L. Ericksen, *Liquid crystals with variable degree of orientation*, Arch. Rational Mech. Anal., **113** (1991), pp. 97–120.

[5] R. Hardt, D. Kinderlehrer and F.H. Lin, *Existence and partial regularity of static liquid crystal configurations*, Comm. Math. Phys., **105** (1986), pp. 547–570.

[6] P.E. Cladis and M. Kléman, *Non-singular disclinations of strength $S = +1$ in nematics*, J. Phys. (Paris), **33** (1972), pp. 591–598.

[7] A. Saupe, *Disclinations and properties of the director field in nematic and cholesteric liquid crystals*, Mol. Cryst. Liq. Cryst., **21** (1973), pp. 211–238.

[8] C. Williams, P.E. Cladis and M. Kléman, *Screw disclinations in nematic samples with cylindrical symmetry*, Mol. Cryst. Liq. Cryst., **21** (1973), pp. 355–373.

[9] P.G. De Gennes, *Phenomenology of short-range-order effects in the isotropic phase of nematic materials*, Phys. Lett., **30 A** (1969), pp. 454–455.

[10] C. Fan, *Disclination lines in liquid crystals*, Phys. Lett., **34 A** (1972), pp. 335–336.

[11] W.J.A. Goossens, *Bulk, interfacial and anchoring energies of liquid crystals*, Mol. Cryst. Liq. Cryst., **124** (1985), pp. 305–331.

[12] E.G. Virga, *Defects in nematic liquid crystals with variable degree of orientation*, *in* Nematics, J.-M. CORON, J.-M. GHIDAGLIA and F. HÉLEIN *Eds.*, Kluwer Academic Publishers, Dordrecht, *etc.* (1991), pp. 371–390.

[13] V.J. Mizel, D. Roccato and E.G. Virga, *A variational problem for nematic liquid crystals with variable degree of orientation*, CMU Research Report, October 1990.

[14] D. Roccato and E.G. Virga, *On surface defects in nematic liquid crystals with variable degree of orientation*, Preprint (1991).

[15] M. Doi, *Molecular dynamics and rheological properties of concentrated solutions of rodlike polymers in isotropic and liquid crystalline phases*, J. Polymer Sci., **19** (1981), pp. 229–243.

[16] E.G. Virga, *Disclinations and hedgehogs in nematic liquid crystals with variable degree of orientation*, Rend. Mat. Acc. Lincei, **1** (1990), pp. 275–280.

[17] L. Ambrosio and E.G. Virga, *A boundary-value problem for nematic liquid crystals with a variable degree of orientation*, to appear in Arch. Rational Mech. Anal. (1991).

[18] R. Hardt, D. Kinderlehrer and F.H. Lin, *Stable defects of minimizers of constrained variational principles*, Ann. Inst. H. Poincaré, Anal. Non Linéaire, **5** (1988), pp. 297–322.